Shwki Al Rashed

Numerical Algorithms in Algebraic Geometry

Shwki Al Rashed

Numerical Algorithms in Algebraic Geometry

Südwestdeutscher Verlag für Hochschulschriften

Impressum/Imprint (nur für Deutschland/only for Germany)
Bibliografische Information der Deutschen Nationalbibliothek: Die Deutsche Nationalbibliothek verzeichnet diese Publikation in der Deutschen Nationalbibliografie; detaillierte bibliografische Daten sind im Internet über http://dnb.d-nb.de abrufbar.
Alle in diesem Buch genannten Marken und Produktnamen unterliegen warenzeichen-, marken- oder patentrechtlichem Schutz bzw. sind Warenzeichen oder eingetragene Warenzeichen der jeweiligen Inhaber. Die Wiedergabe von Marken, Produktnamen, Gebrauchsnamen, Handelsnamen, Warenbezeichnungen u.s.w. in diesem Werk berechtigt auch ohne besondere Kennzeichnung nicht zu der Annahme, dass solche Namen im Sinne der Warenzeichen- und Markenschutzgesetzgebung als frei zu betrachten wären und daher von jedermann benutzt werden dürften.

Coverbild: www.ingimage.com

Verlag: Südwestdeutscher Verlag für Hochschulschriften GmbH & Co. KG
Heinrich-Böcking-Str. 6-8, 66121 Saarbrücken, Deutschland
Telefon +49 681 37 20 271-1, Telefax +49 681 37 20 271-0
Email: info@svh-verlag.de

Approved by: Kaiserslautern, TU, Diss., 2011

Herstellung in Deutschland:
Schaltungsdienst Lange o.H.G., Berlin
Books on Demand GmbH, Norderstedt
Reha GmbH, Saarbrücken
Amazon Distribution GmbH, Leipzig
ISBN: 978-3-8381-1350-0

Imprint (only for USA, GB)
Bibliographic information published by the Deutsche Nationalbibliothek: The Deutsche Nationalbibliothek lists this publication in the Deutsche Nationalbibliografie; detailed bibliographic data are available in the Internet at http://dnb.d-nb.de.
Any brand names and product names mentioned in this book are subject to trademark, brand or patent protection and are trademarks or registered trademarks of their respective holders. The use of brand names, product names, common names, trade names, product descriptions etc. even without a particular marking in this works is in no way to be construed to mean that such names may be regarded as unrestricted in respect of trademark and brand protection legislation and could thus be used by anyone.

Cover image: www.ingimage.com

Publisher: Südwestdeutscher Verlag für Hochschulschriften GmbH & Co. KG
Heinrich-Böcking-Str. 6-8, 66121 Saarbrücken, Germany
Phone +49 681 37 20 271-1, Fax +49 681 37 20 271-0
Email: info@svh-verlag.de

Printed in the U.S.A.
Printed in the U.K. by (see last page)
ISBN: 978-3-8381-1350-0

2

Numerical Algorithms in Algebraic Geometry with Implementation in Computer Algebra System

SINGULAR

Dissertation
by
Shawki Al Rashed

Supervisors
1. Prof. Dr. rer. nat. Gerhard Pfister
2. Prof. Dr. rer. nat. Vladimir Gerdt

University of Kaiserslautern, Germany
Department of Mathematics
Algebra, Geometry and Computer Algebra Group

Vom Fachbereich Mathematik der Technischen Universitt
Kaiserslautern zur Verleihung des akademischen
Grades Doktor der Naturwissenschaften
(Doctor rerum naturalium, Dr. rer. nat.)
genehmigte Dissertation, D 386 .
December 2, 2011

To
My Parents

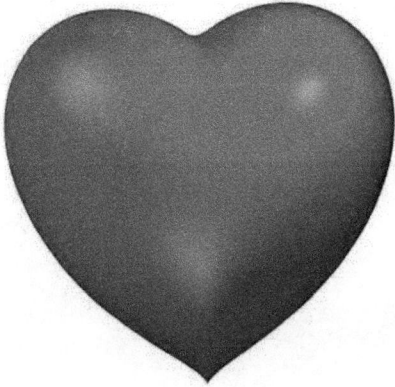

Figure 1: $10(2x^2 + y^2 + z^2 - 1)^3 - x^2 z^3 - 10y^2 z^3 = 0$,
mathematics with heart[1]

---

[1]The heart is displayed by using the SINGULAR library LIB"surf.lib"; (cf. [13]).

Preface

Polynomial systems arise in many applications: robotics, kinematics, chemical kinetics, computer vision, truss design, geometric modeling, and many others. Many polynomial systems have solutions sets, called algebraic varieties, having of multiple irreducible components. A fundamental problem of the numerical algebraic geometry[2] is to decompose such an algebraic variety into its irreducible components. The witness point sets are the natural numerical data structure to encode irreducible algebraic varieties.

The software in the area of numerical algebraic geometry is as young as the area itself. For the purpose of numerical irreducible decomposition there exist at least two software options: BERTINI (cf. [4]) that we use to compute the solutions of the homotopy function and PHCPACK (cf. [37]).

Sommese, Verschelde and Wampler represented the irreducible algebraic decomposition of an affine algebraic variety X as a union of finite disjoint sets $\cup_{i=0}^{d} W_i = \cup_{i=0}^{d} (\cup_{j=1}^{d_i} W_{ij})$ called numerical irreducible decomposition (cf. [25],[26],[28],[29],[30],[32],[33],[34]). The W_i correspond to the pure i-dimensional components, and the W_{ij} present the i-dimensional irreducible components. The numerical irreducible decomposition is implemented in BERTINI (cf. [4]).

We modify this concept using partially Gröbner bases, triangular sets, local dimension, and the so-called zero sum relation. We present in the second chapter the corresponding algorithms and their implementations in SINGULAR (cf. [13]). We give some examples and timings, which show that the modified algorithms are more efficient if the number of variables is not too large. For a large number of variables BERTINI is more efficient.

---

[2]The relationship of numerical algebraic geometry to algebraic geometry is similar to the relationship of numerical linear algebra to linear algebra. It was first proposed by Sommese and Wampler (cf. [34]).

Leykin presented in [18] an algorithm to compute numerically a primary decomposition of an algebraic variety depending on the concept of the deflation of an algebraic variety, and the corresponding algorithms (cf. [25],[26],[28],[29],[30],[33],[34]); they compute the numerical irreducible decomposition.

Depending on the modified algorithm, Algorithm 5 Numerical Irreducible Decomposition in the Chapter 2, we will represent in the third chapter an algorithm and its implementation in SINGULAR to compute the numerical primary decomposition.

The encoding of the irreducible decomposition of algebraic varieties allows us to formulate in the fourth chapter some numerical algebraic algorithms.

Let $X, Y \subset \mathbb{C}^N$ be algebraic varieties defined by the polynomial systems $f = \{f_1, ..., f_n\}$, $g = \{g_1, ..., g_m\} \subset \mathbb{C}[x_1, ..., x_N]$ respectively. Let $\hat{p}$ be the numerical approximate value of a point p on X. The following algorithms:

1. $Incl(X, Y)$ tests if X is a subset of Y.

2. $Equal(X, Y)$ tests whether X is equal to Y.

3. $Degree(X, i)$ computes the degree of a pure i-dimensional component of X.

4. $NumLocalDim(X, \hat{p})$ computes the local dimension of X at p denoted by $dim_p X$.

In the last chapter, Chapter 5, we present two libraries. The first library is used to compute the numerical irreducible decomposition and the numerical primary decomposition of an algebraic variety defined by a polynomial system. The second library contains the procedures of the algorithms in the Chapter 4 to test the inclusion, the equality of two algebraic varieties defined by polynomial systems, and to compute the degree of a pure i-dimensional component, and the local dimension.

Acknowledgement

This page is dedicated to all people, who supported me during my studies.
First I would like to thank my supervisor Prof. Dr. Gerhard Pfister for introducing me to this interesting subject and teaching me the basic knowledge related to my thesis.
I gratefully acknowledge the help, discussion and support of me many colleagues of our group Algebra, Geometry und Computeralgebra (AGAG), especially professor Gert-Martin Greuel, professor Wolfram Decker and professor Thomas Markwig. Their lectures and seminars have been very useful. I also express my thanks to the secretary of our group, Petra Bäsell. She was always willing to help me and my colleagues.
I would like to thank the department of mathematics, especially the Graduate School "Mathematics as a Key Technology", and International School for Graduate Studies (ISGS) for their great assistance given to my studies in Kaiserslautern.

Finally I would especially like to express my gratitude to my wife, Nissren Oudeh, for her love, understanding, continued encouragement and patience throughout all these long years.

vi

Contents

1

List of Algorithms

Chapter 1

General Preliminaries

1.1 Some Definitions and Properties in the Algebraic Geometry

Definition 1.1.1. *Let $<$ be a fixed monomial ordering on $Mon_n = \{x^\alpha, \alpha \in \mathbb{N}^n\}$ and $f \in \mathbb{C}[x_1, ..., x_n]$ written in a unique way as a sum of non-zero terms*

$$f = a_\alpha x^\alpha + a_\beta x^\beta + ... + a_\gamma x^\gamma, \text{ where } x^\alpha > x^\beta > ... > x^\gamma, \text{ and } a_\alpha, a_\beta, ..., a_\gamma \in \mathbb{C}.$$

We define:

1. *$<$ is called a* global ordering *if $1 < x^\alpha$, for all $\alpha \neq (0, 0, ..., 0)$;*

2. *$<$ is called a* local ordering *if $x^\alpha < 1$, for all $\alpha \neq (0, 0, ..., 0)$;*

3. *the* leading monomial *of f, denoted by $LM(f)$, is x^α;*

4. *the* leading exponent *of f, denoted by $LE(f)$, is α;*

5. *the* leading term *of f, denoted by $LT(f)$, is $a_\alpha x^\alpha$.*

Example 1.1.1. *The monomial ordering $<$ is called*

- *lexicographical ordering if and only if*

$$x^\alpha < x^\beta \Leftrightarrow \exists i \in \{1, ..., n\} : \alpha_1 = \beta_1, ..., \alpha_{i-1} = \beta_{i-1}, \text{ and } \alpha_i < \beta_i \ ;$$

- *degree reverse lexicographical ordering if and only if*

$$x^\alpha < x^\beta \Leftrightarrow deg(x^\alpha) < deg(x^\beta) \quad or$$

$$(if\ deg(x^\alpha) = deg(x^\beta) \Rightarrow \exists i \in \{1, ..., n\} : \alpha_n = \beta_n, ..., \alpha_{i+1} = \beta_{i+1}, \text{ and }$$

$$\alpha_i > \beta_i).$$

5

- *negative degree reverse lexicographical ordering if and only if*

$$x^\alpha < x^\beta \Leftrightarrow deg(x^\alpha) > deg(x^\beta) \quad or$$

$$(if \ deg(x^\alpha) = deg(x^\beta) \ and \ \exists i \in \{1, ..., n\} : \alpha_n = \beta_n, ..., \alpha_{i+1} = \beta_{i+1},$$

$$and \ \alpha_i > \beta_i).$$

The lexicographical ordering and the degree reverse lexicographical ordering are global monomial orderings. The negative degree reverse lexicographical ordering is a local monomial ordering.

Definition 1.1.2. *Let I be an ideal defined by the polynomial system $f = \{f_1, ..., f_n\} \subset \mathbb{C}[x_1, ..., x_N]$, and $<$ be a global monomial ordering.*

- *The set of all leading terms of elements of I generates the so-called leading term ideal of I, denoted by $LT(I)$.*

- *A finite set $G = \{g_1, g_2, ..., g_l\} \subseteq I$ is called a Gröbner basis of I if*

$$< LT(g_1), LT(g_2), ..., LT(g_l) >= LT(I).$$

- *A set $T = \{g_1, ..., g_N\}$ of polynomials in $\mathbb{C}[x_1, ..., x_N]$ is called triangular set if for each i*

  1. *$g_i \in \mathbb{C}[x_{N-i+1}, ..., x_N]$,*
  2. *$LM(g_i) = x_{N-i+1}^{m_i}$, for some $m_i > 0$.*

  Hence, g_1 depends only on x_N, g_2 on x_{N-1}, x_N and so on, until g_N which depends on all variables $x_1, ..., x_N$.

Remark 1.1.1. *The triangular sets are used in the* SINGULAR *library "solve.lib" to compute the solutions of a given polynomial system $f = \{f_1, ..., f_n\} \subset \mathbb{C}[x_1, ..., x_N]$, where the system f has only finitely solutions (cf. [13], [14]).*

Theorem 1.1.1. *Let X be a pure n-dimensional algebraic variety in $\mathbb{C}^N$. Then there is a dense Zariski open set U of m-dimensional linear spaces $L \subset \mathbb{C}^N$ such that*

  1. *If $n + m \geq N$, then $L \cap X$ is of dimension $n + m - N$;*

  2. *If $n + m < N$, then $L \cap X$ is empty.*

Proof. (cf. [33])   □

Remark 1.1.2. *The set of all m-dimensional linear subspaces of $\mathbb{C}^N$ is called the Grassmannian and denoted by $G(m, N)$. A general point in $G(m, N)$ is a general slicing plane with respect to the pure n-dimensional component X.*

Definition 1.1.3. *A continuous map $f : X \to Y$ between topological spaces is called proper if $f^{-1}(Z)$ is compact for all $Z \subset Y$ compact.*
An algebraic map $f : X \to Y$ between quasiprojective algebraic sets is called a proper algebraic map if f is proper as a continuous map in the complex topology.

Proposition 1.1.1. *Let $f : X \longrightarrow Y$ be a surjective proper algebraic map between quasiprojective algebraic sets. Assume that X and Y are pure k-dimensional. Then there is a Zariski dense open set $U \subset Y$ such that $f : f^{-1}(U) \longrightarrow U$ is finite.*

Proof. (cf. [15],[33])   □

Definition 1.1.4. *Let $f = \{f_1, ..., f_n\} \subset \mathbb{C}[x_1, ..., x_N]$ be a system of n polynomials on $\mathbb{C}^N$. The **rank** of the polynomial system f, $\mathrm{rank}f$, is defined to be the dimension of the closure of the image $\overline{f(\mathbb{C}^N)} \subset \mathbb{C}^n$.*

Theorem 1.1.2. *Let $f = \{f_1, ..., f_n\} \subset \mathbb{C}[x_1, ..., x_N]$ be a system of n polynomials on $\mathbb{C}^N$. Then all irreducible components of the algebraic variety X defined by the system f have dimension at least equal to $N - \mathrm{rank}f$.*

Proof. (cf. [33])   □

Proposition 1.1.2. *Let $\varphi : X \to Y$ be a holomorphic map between complex spaces X, Y. Then[1] every point $p \in X$ has a neighbourhood $U \subset X$ such that*

$$dim_x X_{\varphi(x)} \leq dim_p X_{\varphi(p)} \ for \ all \ x \in U.$$

Proof. (cf. [15])   □

1.2   Basic Idea of the Homotopy Continuation Method

The homotopy continuation method is used to solve nonlinear equations (cf.[9],[12],[24],[35],[37],[38]). We assume that the number of equations and the number of unknowns are equal. The ideal of the homotopy continuation method is represented as follows.

---

[1]$X_a := \varphi^{-1}(a)$.

- Let

$$F(x_1, ..., x_N) = \begin{pmatrix} F_1(x_1, ..., x_N) \\ . \\ . \\ . \\ F_N(x_1, ..., x_N) \end{pmatrix}$$

  be a system of N polynomials with complex coefficients in N unknowns.

- This method defines a system $G(x) \subset \mathbb{C}[x_1, ..., x_N]$, whose solutions are known. The set of the solutions of $G(x)$ is called a **start solution set** denoted by S_0.

- Define the homotopy function:

$$H : \mathbb{C}^N \times [0, 1] \mapsto \mathbb{C}^N, H(x, t) = (1 - t).F(x) + t.G(x).$$

- $H(x, t)$ defines paths $x(t)$ as t goes from 1 to 0.

- The paths satisfy the Davidenko equation:

$$0 = \frac{dH(x(t), t)}{dt} = \sum_{i=1}^{N} \frac{\partial H}{\partial x_i} \cdot \frac{dx_i}{dt} + \frac{\partial H}{\partial t}.$$

- To compute the paths:

  - Use "ODE" methods to predict.
  - Use Newton's method to correct.

- The system $G(x)$ known as a **start system** is chosen correctly if it has the following properties

  - Triviality: the solutions of $G(x)$ are known;
  - Smoothness: the set of the solutions of $H(x(t), t) = 0$ *for* $0 \leq t \leq 1$ consists of a finite number of smooth paths, each parameterized by t in $(0, 1]$;
  - Accessibility: every isolated solution of $H(x(t), 0) = 0$ can be reached by some path originating at $t = 1$. It follows that this path starts at a solution of $H(x(t), 1) = 0$.

- For every start point $x_0 \in S_0$, we trace a path $c(x(t), t) \in H^{-1}(0)$ from a starting point $(x_0, 1)$ such that $H(x_0, 1) = G(x_0) = 0$ to an endpoint $(x_1, 0)$ as t goes from 1 to 0.

Example 1.2.1. *Let*

$$F(x, y) = \begin{pmatrix} f_1 \\ f_2 \end{pmatrix}$$

be the system of polynomials in $\mathbb{C}^2$, where $f_1 = (y - x + 1)(x - 1)$, $f_2 = (y - x + 1)(y - 2)$.

- *Define the start system $G(x, y)$ as a system of polynomials as follows*

$$G(x, y) = \begin{pmatrix} g_1 \\ g_2 \end{pmatrix},$$

  where $g_1 = x^2 - 4$, $g_2 = y^2 - 4$. The start solution set $S_0 = \{(2, 2), (2, -2), (-2, 2), (-2, -2)\}$ is known.

- *The homotopy function is defined as*

$$H(x, y, t) = (1 - t). \begin{pmatrix} f_1 \\ f_2 \end{pmatrix} + t. \begin{pmatrix} g_1 \\ g_2 \end{pmatrix}.$$

  Where $H(x, y, 1) = G(x, y) = 0$, and $H(x, y, 0) = F(x, y) = 0$.

- *The homotopy continuation method computes the set of the solutions of $F(x, y) = 0$ as t goes from 1 to 0.*

- *Using the computer algebra system* BERTINI *(cf. [4]) we obtain the set of the solutions $S_1 = \{(1, 2), (-0.5, 0.5), (3, 2)\}$ of the system $F(x, y) = 0$.*

1.3 Basic Idea of the Coefficient-Parameter Theory

The using parameter continuation for polynomial systems is that if we can find all solutions to a general member of a family, then we can find all solutions to any other member of that same family.

Theorem 1.3.1. *(Basic Parameter Continuation)* *Let $F(x, q)$ be a system of polynomials in n variables and M parameters,*

$$F(x, q) : \mathbb{C}^n \times \mathbb{C}^M \to \mathbb{C}^n,$$

$$F(x, q) := \{F_1(x, q), ..., F_n(x, q)\},$$

each $F_i(x, q)$ is polynomial in both x and q. Let $N(q)$ denote the number of nonsingular solutions as a function of q:

$$N(q) := \sharp\{x \in \mathbb{C}^n \mid F(x, q) = 0, \ det(\frac{\partial F}{\partial x}(x, q)) \neq 0\}.$$

Then

- *$N(q)$ is finite, and it is the same, say N, for almost all $q \in \mathbb{C}^M$;*

- *For all $q \in \mathbb{C}^M$, $N(q) \leq N$;*

- *The subset of $\mathbb{C}^M$ where $N(q) = N$ is a Zariski open set. That is, the exceptional set $Q := \{q \in \mathbb{C}^n \mid N(q) < N\}$ is an affine algebraic variety contained within an algebraic variety of dimension n-1;*

- *The homotopy function $F(x, \lambda(t)) = 0$ with $\lambda(t) : [0, 1] \to \mathbb{C}^M \setminus Q$ has N continuous nonsingular solution paths $x(t) \in \mathbb{C}^n$;*

- *As t goes from 1 to 0, the limits of the solution paths of the homotopy function $F(x, \lambda(t)) = 0$ with $\lambda(t) : [0, 1] \to \mathbb{C}^M$ and $\lambda(t)$ not in Q for $t \in (0, 1]$ include all the nonsingular roots of $F(x, \lambda(0)) = 0$.*

Proof. (cf. [33])   □

Lemma 1.3.1. *Fix a point $q_0 \in \mathbb{C}^n$ and a proper algebraic set $A \subset \mathbb{C}^n$. For almost all $q_1 \in \mathbb{C}^n$, the one-real-dimensional open line segment*

$$L(t) := tq_1 + (1 - t)q_0, \quad t \in (0, 1],$$

is contained in $\mathbb{C}^n \setminus A$.

Proof. (cf. [33])   □

Remark 1.3.1. *The lemma 1.3.1 and item 5 of the theorem 1.3.1 above imply that for a given target set of parameters q_0 almost any starting set of parameters q_1 will give a homotopy function $F(x, tq_1 + (1 - t)q_0) = 0$ whose solution paths include all the nonsingular solutions of $F(x, q_0) = 0$ at their endpoints as t goes from 1 to 0 on the real line.*

Example 1.3.1. *Let $l(t) = tq_1 + (1 - t)q_0$ be the one-real dimensional open line segment. Where the linear space L_0 defined by the linear polynomial $l_0 = x + 3y - 2$ is the target set of parameters q_0 , and the linear space L_1 defined by the linear polynomial $l_1 = x + y - 1$ is the starting set of* *parameters q_1 as t goes from 1 to 0. Then we can define $F(x, y, t)$ as a system of polynomials in 2 variables x, y and one parameter t,*

$$F(x, y, t) : \mathbb{C}^2 \times \mathbb{C} \rightarrow \mathbb{C}^2,$$

$$F(x, y, t) = \left(\begin{array}{c} (x^2 + y^2 - 5)(x - y) \\ tl_1 + (1 - t)l_0 \end{array} \right) = \left(\begin{array}{c} (x^2 + y^2 - 5)(x - y) \\ x + (3 - 2t)y + t - 2 \end{array} \right).$$

Particularly, $F(x, y, t)$ is a homotopy function written as follows

$$F(x, y, t) = t. \left(\begin{array}{c} (x^2 + y^2 - 5)(x - y) \\ x + y - 1 \end{array} \right) + (1 - t). \left(\begin{array}{c} (x^2 + y^2 - 5)(x - y) \\ x + 3y - 2 \end{array} \right).$$

with the start system $\{(x^2 + y^2 - 5)(x + y), x - y + 1\}$ and the start solution set $\{(2, -1), (-1, 2), (\frac{1}{2}, \frac{1}{2})\}$. The set of the solutions $V(F(x, y, q_0)) = \{(2, 2346, -0.0782), (-1.8346, 1.2782), (0.5, 0.5)\}$ is computed using the computer algebra system BERTINI (cf. [4]) as t goes from 1 to 0 on the one-real dimensional line $l(t)$.

1.4 Basic Idea of the Monodromy Action on an Algebraic Variety

Definition 1.4.1. *Let X be a topological space.*

- *Let x_0 be a point of X and $f : [0, 1] \longrightarrow X$ be a continuous function. f is called loop with base x_0 if it has the property $f(0) = f(1) = x_0$.*

- *Any two loops, say f and g, are considered equivalent if there is a continuous function $h : [0, 1] \times [0, 1] \longrightarrow X$ with property that for all $t \in [0, 1]$, h(t,0)=f(t), h(t,1)=g(t) and $h(0, t) = x_0 = h(1, t)$. Such a function h is called a homotopy from f to g, and the corresponding classes are called homotopy classes.*

- *The set of all homotopy classes of loops with base point x_0 forms the fundamental group of X at point x_0, which is denoted by $\Pi_1(X, x_0)$.*

Monodromy Action on an Algebraic Variety

Let $X \subset \mathbb{C}^N$ be a pure k-dimensional algebraic variety, and $\overline{G}(N - k, N)$ denote the Grassmannian of linear spaces in the projective space $\mathbb{P}^N$. We close X to get a pure k-dimensional projective algebraic variety $\overline{X} \subset \mathbb{P}^N$. Let $R := \{(L_{N-k}, x) \in \overline{G}(N - k, N) \times \overline{X} \mid x \in L_{N-k} \cap \overline{X}\}$ be the family of the intersections $L_{N-k} \cap \overline{X}$ for k-dimensional linear spaces $L_{N-k} \subset \mathbb{P}^N$, which it is a projective algebraic set. Then we have two maps $p : R \longrightarrow \overline{G}(N-k, N)$ and $q : R \longrightarrow \overline{X}$ induced by the product projections on $\overline{G}(N-k, N) \times \overline{X}$. A generic linear space L_{N-k} intersects $\overline{X}$ in a set of d distinct points, where d is the degree of X. From "Proposition 1.1.1" there is a Zariski open dense set $U \subset \overline{G}(N-k, N)$ such that $p_{p^{-1}(U)} : p^{-1}(U) \longrightarrow U$ is a finite covering. Fix a general point in U, say L as base point, then we have the monodromy action of the fundamental group $\Pi_1(U, L)$ on the set $p^{-1}(L)$.

Lemma 1.4.1. *If X_i is an irreducible component of X, then the above monodromy action acts transitively on the set $X_i \cap p^{-1}(L)$.*

Proof. (cf. [33])   □

Example 1.4.1. *Let X be the algebraic variety of dimension one in $\mathbb{C}^2$ defined by the polynomial $f(x, y) = (x^2 + y^2 - 5)(x - 2y - 3)$. Let $L_1 \subset \mathbb{C}^2$ be the linear space of dimension one defined by the linear polynomial $l_1 = x + y - 3$. Define the homotopy function :*

$$h(t, x(t), y(t)) = \begin{pmatrix} \alpha(t) \\ f(x(t), y(t)) \end{pmatrix},$$

where $\alpha : [0, 1] \longrightarrow p^{-1}(L)$ is given by

$$\alpha(t) = (1-t)l_0 + t l_1 = x(t) + y(t) - 2t - 1.$$

Let L_0 be the 1-dimensional linear space defined by the polynomial $l_0 = x + y - 1$. Then $\alpha(t)$ moves on X and maps a point in $L_1 \cap X$ to a point in $L_0 \cap X$ as t goes from 1 to 0.

Definition 1.4.2. *A linear projection $\pi : \mathbb{C}^N \longrightarrow \mathbb{C}^M$ is a surjective affine map given by*
$\pi(x_1, ..., x_N) = (L_1(x_1, ..., x_N), ..., L_M(x_1, ..., x_N))$, *where $M \leq N$ and*

$$L_i(x_1, ..., x_N) := a_{i0} + \sum_{j=1}^{N} a_{ij} x_j.$$

for $i = 1, ..., N$, $j = 0, 1, ..., M$ and $a_{ij} \in \mathbb{C}$. π is called a generic linear projection if the coefficients a_{ij} are chosen randomly.

Proposition 1.4.1. *Let X be a reduced pure k-dimensional algebraic set in $\mathbb{C}^N$, and $X = \cup_{i=1}^{r} X_i$ be an irreducible decomposition of X. Let $\pi : \mathbb{C}^N \longrightarrow \mathbb{C}^k$ denote a generic linear projection, and $x, y \in \mathbb{C}^k$ be general points, with*

$L := \pi^{-1}(x) \subset \mathbb{C}^N$ *a general linear space of dimension* $N - k$ *intersecting* X
in the set $W := X \cap L$.

For all $i = 1, ..., r$, *set* $W_i := X_i \cap L$ *and let* d_i *be the cardinality of the set*
W_i. *Let* d *denote the cardinality of* W, *i.e*, $d := \sum_{i=1}^{r} d_i$.

Let U *denote the Zariski open set of the line* $l \subset \mathbb{C}^k$ *containing* x, y, *consisting*
of the $u \in l$ *such that* $\pi^{-1}(u)$ *is transverse to* X. *Let* $Sym(W)$ *(respectively*
$Sym(W_i))$ *denote the symmetric group of all permutations of* W *(respectively*
$W_i)$. *Considering* L *as a base point of* U, *then the image in* $Sym(W)$ *of*
the monodromy action of the fundamental group $\Pi_1(U, L)$ *on* W *is the direct*
sum $\oplus_{i=1}^{r} Sym(W_i)$.

Proof. (cf. [25])   □

Chapter 2

Numerical Irreducible Decomposition of an Algebraic Variety

The algebraic irreducible decomposition of an affine algebraic variety is represented as a union of finite disjoint sets $W = \cup_{i=0}^{d} W_i = \cup_{i=0}^{d}(\cup_{j=1}^{d_i} W_{ij})$ called numerical irreducible decomposition denoted by $"N.I.D"$ (cf. [25],[26],[28],[29],[30],[32],[33],[34]).

Modifying this concept by using partially Gröbner bases, triangular sets, local dimension, and the so-called "zero sum relation" we present in this chapter modified algorithms and their implementation in SINGULAR (cf. [13]) to compute the numerical irreducible decomposition. We will give some examples and timings, which show that the modified algorithms are more efficient if the number of variables is not too large. For a large number of variables BERTINI (cf. [4]) is more efficient[1].

Definition of $"N.I.D"$

Given a system of n polynomials in the polynomial ring $\mathbb{C}[x_1, ..., x_N]$,

$$f(x_1, ..., x_N) = \begin{pmatrix} f_1(x_1, ..., x_N) \\ \cdot \\ \cdot \\ \cdot \\ f_n(x_1, ..., x_N) \end{pmatrix}. \tag{2.1}$$

---

[1]Note that each step of the numerical irreducible decomposition is parallelizable. For our comparisons we did not use the parallel version of BERTINI resp. the parallel version of SINGULAR .

Let $X = V(f) \subset \mathbb{C}^N$ be the algebraic variety defined by the system above. X has a unique algebraic decomposition into d pure i-dimensional components X_i, $X = \cup_{i=0}^d X_i$. Where $X_i = \cup_{j_i}^{d_i} X_{ij}$ is empty or the union of d_i i-dimensional irreducible components.

Definition 2.0.3. *The union $X = \cup_{i=0}^d X_i$ is called the algebraic irreducible decomposition of X. Here d is the dimension of X, $X_i := \cup_{j \in J_i} X_{ij}$ is a finite union of i-dimensional irreducible components of X, called the pure i-dimensional component of X.*

Definition 2.0.4. *The numerical irreducible decomposition denoted by N.I.D of X is defined as a union of finite disjoint sets $W = \cup_{i=0}^d W_i = \cup_{i=0}^d (\cup_{j=1}^{d_i} W_{ij})$. The W_i are called the i-witness point sets of the pure i-dimensional components X_i of X, and the finite sets W_{ij} are called irreducible i-witness point sets of irreducible components X_{ij} of dimension i. Where the irreducible witness point sets have the following properties:*

1. *$W_{ij} \subset X_{ij}$,*

2. *$\sharp(W_{ij}) = deg(X_{ij})$,*

3. *$W_{ij} \cap W_{il} = \emptyset$ for $j \neq l$.*

Example 2.0.2. *Let X be the algebraic variety defined by the polynomial system*

$$f(x,y,z) = \begin{pmatrix} (x^2 + y^2 + z^2 - 6)(x-y)(x-1) \\ (x^2 + y^2 + z^2 - 6)(x-z)(y-2) \\ (x^2 + y^2 + z^2 - 6)(x-y)(x-z)(z-3) \end{pmatrix}.$$

The algebraic irreducible decomposition of X is presented as the union $X = \cup_{i=0}^2 X_i$, where X_i is a union of i-dimensional irreducible components of X as follows.

- *$X_0 := X_{01}$, where $X_{01} = \{(1,2,3)\}$,*

- *$X_1 := X_{11} \cup X_{12} \cup X_{13}$, where $X_{11} = \{(2,2,z) \mid z \in \mathbb{C}\}$, $X_{12} = \{(1,y,1) \mid y \in \mathbb{C}\}$, and $X_{13} = \{(x,x,x) \mid x \in \mathbb{C}\}$,*

- *$X_2 := X_{21}$, where $X_{21} = \{(x,y,z) \in \mathbb{C}^N \mid x^2 + y^2 + z^2 = 6\}$.*

The numerical irreducible decomposition of X is given as the union $W = \cup_{i=0}^2 W_i$, where W_i is a union of irreducible i-witness point sets as follows.

- *$W_0 := W_{01}$, where $W_{01} = \{(1,2,3)\}$ represents X_{01},*

- $W_1 := W_{11} \cup W_{12} \cup W_{13}$, where $W_{11} = \{(2,2,-2)\}$, $W_{12} = \{(1,0,1)\}$, and $W_{13} = \{(\frac{2}{3}, \frac{2}{3}, \frac{2}{3})\}$ represent X_{11}, X_{12}, and X_{13} respectively,

- $W_2 = W_{21}$, where $W_{21} = \{(-1,2,1),(1,2,-1)\}$ represents X_{21}.

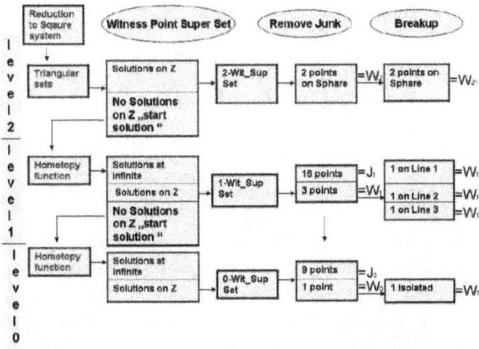

Figure 2.1: Flow Diagram of the Example 2.0.2.

The numerical irreducible decomposition proceeds in four steps:

1. The first step reduces the polynomial system to a system of N polynomials in N variables (cf. [26],[33],[34]).

2. The second step computes a finite set $\widehat{W}_i$ called witness point super set for each non-empty pure i-dimensional component X_i. $\widehat{W}_i$ consists of points of X_i and J_i a set of points on components of larger dimension, the so-called junk point set (cf. [26],[28],[33]).

3. The third step removes the points of J_i from $\widehat{W}_i$ to obtain a subset W_i of the pure i-dimensional component X_i (cf. [33]).

4. The fourth step breakups W_i into irreducible witness point sets representing the i-dimensional irreducible components of X (cf. [25],[32]).

In the first section of this chapter we reduce the polynomial system to a system of N polynomials in N variables (cf. [26],[33],[34]). In [26],[28],[33] the cascade algorithm is used to compute the witness point super sets $\widehat{W}_i$. In the second section of this chapter we modify this algorithm replacing the use of the homotopy function by Gröbner basis computations at certain levels.

In [33] the parameter continuation for polynomial systems is used to remove junk points from $\widehat{W}_i$ to obtain the i-witness point set W_i. In the third section of this chapter we give a modified algorithm using local dimension, Gröbner bases in the zero-dimensional case, and the homotopy function to compute the i-witness point set W_i.

The breakup of the witness point set W_i into irreducible witness point sets is achieved using two algorithms (cf. [25],[32]). The first algorithm computes the points on the same irreducible component in the witness point set using path tracking techniques. The second algorithm computes a linear trace for each component which certifies the decomposition. In the fourth section of this chapter we explain how to use the zero sum relation (cf. [10]) and the monodromy action on the algebraic variety to breakup W_i into irreducible witness point sets. In the last section of this chapter we give examples and timings to compare the implementations of BERTINI and SINGULAR .

2.1 Reduction to the Square System

The computation of the numerical irreducible decomposition uses numerical homotopy continuation methods . It is usually convenient, and sometimes requires that the number n of polynomials of a given polynomial system is equal to the number N of variables (cf. [36],[37],[33]). Therefore we reduce the polynomial system $f(x)$, (2.1), to a system of N polynomials in N variables which is known as a square system of $f(x)$.

Let $\Lambda \in \mathbb{C}^{N \times n}$ be a generic $N \times n$ matrix. We will form a square system

$$\Lambda.f(x) = \begin{pmatrix} \lambda_{11}f_1(x) + ... + \lambda_{1n}f_n(x) \\ \cdot \\ \cdot \\ \cdot \\ \lambda_{N1}f_1(x) + ... + \lambda_{Nn}f_n(x) \end{pmatrix}, \tag{2.2}$$

of the polynomial system $f(x)$, (2.1), as follows.

If $N = n$, then Λ is invertible, the system $f(x) = \Lambda^{-1}\Lambda.f(x) = 0$

and the system $\Lambda.f(x) = 0$ are equivalent.

If $N > n$, we may break $\Lambda \in \mathbb{C}^{N\times n}$ into two submatrices by rows; the matrix Λ_1 formed from the first n rows is an invertible matrix. Let Λ_2 be the remaining $(N-n) \times n$ matrix formed from the last $N-n$ rows of Λ. Define

$$\Gamma = \begin{pmatrix} \Lambda_1^{-1} & 0_{n\times(N-n)} \\ -\Lambda_2.\Lambda_1^{-1} & I_{N-n} \end{pmatrix}$$

to be a matrix in $\mathbb{C}^{N\times n}$. Γ is invertible and $\Lambda.f(x) = 0$ is equivalent to

$$\Gamma.\Lambda.f(x) = \begin{pmatrix} \Lambda_1^{-1} & 0_{n\times(N-n)} \\ -\Lambda_2.\Lambda_1^{-1} & I_{N-n} \end{pmatrix} \begin{pmatrix} \Lambda_1 \\ \Lambda_2 \end{pmatrix}. \begin{pmatrix} f_1(x) \\ \cdot \\ \cdot \\ \cdot \\ f_n(x) \end{pmatrix} =$$

$$\begin{pmatrix} \mathbb{I}_{N\times 1} \\ 0_{(N-1)\times 1} \end{pmatrix}. \begin{pmatrix} f_1(x) \\ \cdot \\ \cdot \\ f_n(x) \end{pmatrix} = \begin{pmatrix} f_1(x) \\ \cdot \\ \cdot \\ f_n(x) \\ 0_{(N-1)\times 1} \end{pmatrix} = 0.$$

If $N < n$, we break Λ into two submatrices as $\Lambda = \begin{pmatrix} \Lambda_1 & \Lambda_2 \end{pmatrix}$. Where Λ_1 is invertible $N \times N$ submatrix of Λ and Λ_2 is a $N \times (n-N)$ submatrix of Λ. The system $\Lambda.f(x) = 0$ is equivalent to the system

$$\Lambda_1^{-1}\Lambda.f(x) = \Lambda_1^{-1} \begin{pmatrix} \Lambda_1 & \Lambda_2 \end{pmatrix}. \begin{pmatrix} f_1(x) \\ \cdot \\ \cdot \\ \cdot \\ f_n(x) \end{pmatrix} =$$

$$= \begin{pmatrix} f_1(x) \\ \cdot \\ \cdot \\ f_N(x) \end{pmatrix} + \Lambda_1^{-1}\Lambda_2. \begin{pmatrix} f_{N+1}(x) \\ \cdot \\ \cdot \\ f_n(x) \end{pmatrix} = \begin{pmatrix} \mathbb{I}_{N\times N} & \tilde{\Lambda} \end{pmatrix}.f(x) = 0;$$

where $\tilde{\Lambda} := \Lambda_1^{-1}\Lambda_2$ is $N \times (n-N)$ matrix.
We note that only in case $N < n$ this construction interesting.

The following theorem explains the relation between the irreducible components of the algebraic variety $X = V(f)$ defined by the system (2.1) and irreducible components of the algebraic variety $V(\Lambda.f(x))$ defined by the system (2.2).

Theorem 2.1.1. *Let*

$$f(x) = \begin{pmatrix} f_1(x_1, ..., x_N) \\ \cdot \\ \cdot \\ \cdot \\ f_n(x_1, ..., x_N) \end{pmatrix} = 0$$

be a system of polynomials on $\mathbb{C}^N$ and $k \leq N$ be an integer. Assume that $X \subset \mathbb{C}^N$ is an irreducible affine algebraic set. Then there is a nonempty Zarisiki open set U of $k \times n$ matrices $\Lambda \in \mathbb{C}^{k \times n}$ such that for $\Lambda \in U$ the following hold:

- *if $\dim X > N - k$, then X is an irreducible component of $V(f)$ if and only if it is an irreducible component of $V(\Lambda.f)$;*

- *if $\dim X = N - k$, then X is an irreducible component of $V(f)$ implies that X is also an irreducible component of $V(\Lambda.f)$;*

- *if X is an irreducible component of $V(f)$, its multiplicity as a component of $\Lambda.f(x)=0$ is greater than equal to its multiplicity as a component of $f(x) = 0$, with equality if either multiplicity is one.*

Proof. (cf. [26],[33],[34]). $\qquad\qquad\qquad\qquad\qquad\qquad\qquad\qquad\qquad\quad$ $\square$

Remark 2.1.1. *Theorem 2.1.1 shows that*

- *the positive dimensional irreducible components of $V(f)$ are the same as the positive dimensional irreducible components of $V(\Lambda.f)$;*

- *the zero dimensional irreducible components of $V(f)$ are contained in the set of the isolated components of irreducible components of $V(\Lambda.f)$.*

Example 2.1.1. *Let $Z = V(F) \subset \mathbb{C}^2$ be an algebraic variety defined by the system $F = \{(x - 1)(x + y + 1), (x - 1)xy, (x - 1)y\}$ of 3 polynomials in 2 variables. We note that Z is decomposed in the 1-dimensional irreducible component $Z_1 = V(x-1)$ and zero dimensional component $Z_0 = V(x+1, y)$. The square system f of the system F is given as follows*

$$f = \begin{pmatrix} (x - 1)(x + y + 1) \\ (x - 1)xy \end{pmatrix} + \begin{pmatrix} 1 \\ 2 \end{pmatrix} . (\; (x - 1)y \;) =$$

$$= \left(\begin{array}{c} (x-1)((x+2y+1) \\ y(x-1)(x+2) \end{array} \right).$$

Let $X = V(f)$ be the algebraic variety defined by the square system $f = \{(x-1)((x+2y+1), y(x-1)(x+2)\}.$

Then X is decomposed into 1-dimensional irreducible $X_1 = V(x-1)$ and zero dimensional components $X_{01} = V(x+1, y), X_{02} = V(x+2, 2y-1).$

Listing 1 REDUCTION TO SQUARE SYSTEM: *Re2SquaSys(F)*

Input: $F = \{F_1, ..., F_n\} \subset \mathbb{C}[x_1, ..., x_N]$ system of n polynomials in N variables.

Output: $f = \{f_1, ..., f_N\} \subset \mathbb{C}[x_1, ..., x_N]$ square system of N polynomials in N variables.

  if $N = n$ **then**
    set $f_1 := F_1, ..., f_N = F_N$;
  else
    if $N > n$ **then**
      for $i = 1$ to N **do**
        if $i \leq n$ **then**
          $f_i := F_i$;
        else
          $f_i := 0$;
    else
      for $i = 1$ to N **do**
        set $f_i := f_i + \sum_{j=N+1}^{n} \lambda_{ij} f_j$, where $\lambda_{ij} \in \mathbb{C}$ are generic;
  return $f = \{f_1, ..., f_N\}$;

SINGULAR Example:

```
LIB"NumerDecom.lib";
ring r= 0,(x,y,z),dp;
poly f1= x^2+ y+z;
poly f2= xy+yz;
ideal I=f1, f2;
def S=re2squ(I);
setring S;
J;
  ==> J[1]= x^2+ y+z;
```

```
==> J[2]= xy+yz;
==> J[3]= 0;
ideal I=J,x+y+z-2, xy-z;
def S=re2squ(I);
setring S;
J;
==> J[1]= y^2+y
==> J[2]= xy+x
==> J[3]= x^2+y
```

2.2 Witness Point Super Set

We do not know an algorithm for directly computing the i-witness point set W_i of a pure i-dimensional component X_i of the algebraic variety X. Therefore we present in this section an algorithm to compute a larger set $\widehat{W}_i$ that contains W_i known as a witness point super set of a pure i-dimensional component X_i (cf. [33]).

Definition 2.2.1. *Let X be an affine algebraic variety in $\mathbb{C}^N$ defined by the polynomial system $f(x)$ of N polynomials with N unknowns and X_i be a pure i-dimensional component of X. Let L_i be a generic linear space in $\mathbb{C}^N$ of dimension $N - i$. A finite set $\widehat{W}_i \subset \mathbb{C}^N$ is called i-witness point super set of X_i if*

$$X_i \cap L_i \subset \widehat{W}_i \subset X \cap L_i.$$

The union $\widehat{W}$ of all i-witness point super set is called a witness point super set for X.

The computation of the witness point super set is based on the following idea. A generic linear space L_i of dimension $N - i$ defined by i generic linear polynomials will intersect a pure i-dimensional component in finite set of points W_i and meet pure j-dimensional components of the algebraic variety X in some points for $j = i + 1, ..., N$. We can choose the L_i such that

$$L_0 := \mathbb{C}^N \supset L_1 \supset ... \supset L_{i-1} \supset L_i \supset ... \supset L_{N-1}.$$

In [26],[33] an embedding of the system $f(x)$ into a family of systems of polynomials depending on $2N$ variables $(x, z) \in \mathbb{C}^{2N}$ and a large space of parameters is given. Then one singles out $N + 1$ of the systems, $\Omega_i(x, z)$ for i from N to 0 choosing particular values of the parameters and a homotopy function H_i going from a start system Ω_i to Ω_{i-1} from $i = N - 1$ to $i = 1$, using the homotopy continuation method (cf. [2],[21],[36],[37]).

Using Gröbner bases we will show that the modified algorithm below starts with $d \leq N - 1$ and does not need to define a start system, whose solutions are known, for the homotopy function H_i going from a start system Ω_i to Ω_{i-1}:

- The algebraic variety X is defined by the system of N polynomials

$$f(x) = \begin{pmatrix} f_1(x_1, .., x_N) \\ \cdot \\ \cdot \\ f_N(x_1, .., x_N) \end{pmatrix} \ on \ \mathbb{C}^N.$$

- The generic linear space L_i is defined by i linear equation

$$l_j := a_j + a_{j1}x_1 + ... + a_{jN}x_N.$$

for $j = 1, ..., i$ and $a_j, a_{j1}, .., a_{jN} \in \mathbb{C}$ are randomly chosen.

- Fix linear coordinates $z_1, z_2, ..., z_N$ on $\mathbb{C}^N$.
  Define the family $\Omega_i(f)$ for $i = 0, 1, ..., N$ as follows:

  - $\Omega_0(f)(x, z) := f(x)$.
  - For $i > 0$, choose λ_{kj} at random in $\mathbb{C}$, where $k = 1, ..., N$, $j = 1, ..., i$.

  - Define $\Omega_i(f)$ as a system of polynomials

$$\Omega_i(f)(x, z_1, ..., z_i) =: \begin{pmatrix} f_1(x) + \sum_{j=1}^{i} \lambda_{1j} z_j \\ \cdot \\ \cdot \\ f_N(x) + \sum_{j=1}^{i} \lambda_{Nj} z_j \\ l_1 + z_1 \\ \cdot \\ \cdot \\ l_i + z_i \end{pmatrix}$$

  on $\mathbb{C}^{N \times i}$.
  - We note that the solutions $(x, z_1, ..., z_i) \in \mathbb{C}^N \times \mathbb{C}^i$ of $\Omega_i(f)(x, z_1, ..., z_i) = 0$ are naturally identified with the solutions $(x, z_1, .., z_i, 0, .., 0) \in \mathbb{C}^N \times \mathbb{C}^N$ of the system $\Omega_i(f)(x, z_1, ..., z_d) = 0$.

- We note that the $\Omega_i(f)(x, z_1, ..., z_i)$ depends on the choice of the parameters

$$
\begin{pmatrix}
a_1 & a_{11} & . & . & a_{1N} & \lambda_{11} & . & . & \lambda_{N1} \\
a_2 & a_{21} & . & . & a_{2N} & \lambda_{12} & . & . & \lambda_{N2} \\
. & . & & & . & . & & & . \\
. & . & & & . & . & & & . \\
a_i & a_{i1} & . & . & a_{iN} & \lambda_{1i} & . & . & \lambda_{Ni}
\end{pmatrix}
\in \mathbb{C}^{i \times (N+1)} \times \mathbb{C}^{i \times N}.
$$

(2.3)

Example 2.2.1. • *Let*

$$
f(x, y, z) =
\begin{pmatrix}
(x^2 + y^2 + z^2 - 4)(x - y)(x - 1) \\
(x^2 + y^2 + z^2 - 4)(x - z)(y - 1) \\
(x^2 + y^2 + z^2 - 4)(x - y)(x - z)(z - 1)
\end{pmatrix}
$$

be a system of polynomial in $\mathbb{C}[x, y, z]$.

- *Define a generic linear space* $V(L_2) \subset \mathbb{C}^3$ *defined by the system*

$$
L_2 = \begin{pmatrix} l_1 \\ l_2 \end{pmatrix}, \quad \text{where } l_1 = 1 + x + y + z, \ l_2 = 1 + 2x + y + z.
$$

- *Choose* $\lambda_{11} = \lambda_{12} = 1, \lambda_{21} = 2, \lambda_{22} = 1, \lambda_{31} = 1, \lambda_{32} = 3$.
  Fix linear coordinates z_1, z_2 *on* $\mathbb{C}^2$.

- *The specified set of parameters*

$$
Y_2 := \begin{pmatrix} 1 & 1 & 1 & 1 & 1 & 2 & 1 \\ 1 & 2 & 1 & 1 & 1 & 1 & 3 \end{pmatrix} \in \mathbb{C}^{2 \times 4} \times \mathbb{C}^{2 \times 3}.
$$

•

$$
\Omega_2(f) :=
\begin{pmatrix}
(x^2 + y^2 + z^2 - 4)(x - y)(x - 1) + z_1 + z_2 \\
(x^2 + y^2 + z^2 - 4)(x - z)(y - 1) + 2z_1 + z_2 \\
(x^2 + y^2 + z^2 - 4)(x - y)(x - z)(z - 1) + z_1 + 3z_2 \\
1 + x + y + z + z_1 \\
1 + 2x + y + z + z_2
\end{pmatrix}.
$$

- *Define a generic Linear space* $V(L_1) \subset \mathbb{C}^3$ *defined by the system* L_1 *of the linear polynomial* l_1. *The specified set of parameters*

$$
Y_1 = \begin{pmatrix} 1 & 1 & 1 & 1 & 1 & 2 & 1 \end{pmatrix} \in \mathbb{C}^{1 \times 4} \times \mathbb{C}^{1 \times 3}.
$$

-

$$\Omega_1(f) := \begin{pmatrix} (x^2+y^2+z^2-4)(x-y)(x-1)+z_1 \\ (x^2+y^2+z^2-4)(x-z)(y-1)+2z_1 \\ (x^2+y^2+z^2-4)(x-y)(x-z)(z-1)+z_1 \\ 1+x+y+z+z_1 \end{pmatrix}.$$

-

$$\Omega_0(f) := \begin{pmatrix} (x^2+y^2+z^2-4)(x-y)(x-1) \\ (x^2+y^2+z^2-4)(x-z)(y-1) \\ (x^2+y^2+z^2-4)(x-y)(x-z)(z-1) \end{pmatrix}.$$

Lemma 2.2.1. *There is a nonempty Zariski open set U of points*

$$\begin{pmatrix} a_1 & a_{11} & . & . & a_{1N} & \lambda_{11} & . & . & \lambda_{N1} \\ a_2 & a_{21} & . & . & a_{2N} & \lambda_{12} & . & . & \lambda_{N2} \\ . & . & & & . & . & & & . \\ . & . & & & . & . & & & . \\ . & . & & & . & . & & & . \\ a_N & a_{N1} & . & . & a_{NN} & \lambda_{1N} & . & . & \lambda_{NN} \end{pmatrix} \in \mathbb{C}^{N\times(N+1)} \times \mathbb{C}^{N\times N}$$

such that for each $i = 1, ..., N$:

- *The solutions of $\Omega_i(f)(x, z_1, ..., z_i) = 0$ with $(z_1, ..., z_i) \neq 0$ are isolated and nonsingular.*

- *Let X be an irreducible component of $Z = V(f) \subset \mathbb{C}^N$ of dimension i, the set of isolated solutions of $\Omega_i(f)(x, z_1, ..., z_i) = 0$, with $(z_1, ..., z_i) = 0$, contains $\deg(X_{red})$ generic points of X_{red}, where X_{red} is the reduction of X.*

- *The solutions of $\Omega_i(f)(x, z_1, ..., z_i) = 0$ with $(z_1, ..., z_i) \neq 0$ are the same as the solutions of $\Omega_i(f)(x, z_1, ..., z_i) = 0$ with $z_i \neq 0$.*

Proof. (cf. [26],[33]). □

So far this is the approach which can be found in [26],[33] to compute the witness point super sets. Now we give some modifications.

- Compute $d := dim(X)$ to be the dimension of X using Gröbner bases (cf.[13],[17]).

- Since X is of dimension d, then X has no component of dimension greater than d, i.e. $L_i \cap X = \emptyset$ for $i > d$. Therefore the modified algorithm below starts at $d \leq N - 1$.

- We note that the algebraic set $T_d := V(\Omega_d(f)(x, z_1, ..., z_d)) \subset \mathbb{C}^{N \times d}$ is 0-dimensional. Then T_d is computed using a solver based on triangular sets (cf.[13],[17]).

- Breakup the set T_d into two sets

$$S_d := \{(x_1, ..., x_N, z_1, ..., z_d) \in T_d \mid \exists j \in \{1, ..., d\} \text{ such that } z_j \neq 0\},$$

$$\widehat{W}_d := \{(x_1, ..., x_N) \in \mathbb{C}^N \mid (x_1, ..., x_N, z_1, ..., z_d) \in T_d \setminus S_d\}.$$

- For $i = d - 1, ..., 0$

  - Define the homotopy function

  $$H_i(x, z_1, ..., z_i, t) = t.\Omega_{i+1}(f) + (1 - t). \begin{pmatrix} \Omega_i(f) \\ z_{i+1} \end{pmatrix}.$$

  - Compute $T_i := V(H_i(x, z_1, ..., z_i, t)) \subset \mathbb{C}^{N \times i}$ using the homotopy continuation method with the start system Ω_{i+1} and the start solution set S_{i+1} as t goes from 1 to 0.

  - Breakup the set T_i into two sets

  $$S_i := \{(x_1, ..., x_N, z_1, ..., z_i) \in T_i \mid \exists j \in \{1, ..., i\} \text{ such that } z_j \neq 0\},$$

  $$\widehat{W}_i := \{(x_1, ..., x_N) \in \mathbb{C}^N \mid (x_1, ..., x_N, z_1, ..., z_i) \in T_i \setminus S_i\}.$$

We will illustrate our modifications of the Algorithm 2 by an example.

Example 2.2.2. *Let $X \subset \mathbb{C}^3$ be the algebraic variety defined by the polynomial system*

$$f(x, y, z) = \begin{pmatrix} (x^3 + z)(x^2 - y) \\ (x^3 + y)(x^2 - z) \\ (x^3 + z)(x^3 + y)(z^2 - y) \end{pmatrix}.$$

The dimension of X is 1.
The algorithm in [26],[33] starts at level 2, while the modified algorithm starts at level 1 to compute the witness point super set $\widehat{W}$ for the algebraic variety X as follows.

- $L = \{l_1\}$ *the set of 1 generic linear polynomials, where $l_1 = x + y + z - 1$;*

-

$$\Omega_0(f)(x, y, z) := f(x, y, z) = \begin{pmatrix} (x^3 + z)(x^2 - y) \\ (x^3 + y)(x^2 - z) \\ (x^3 + z)(x^3 + y)(z^2 - y) \end{pmatrix};$$

- $\lambda_{11} := 1, \lambda_{12} := 5, \lambda_{13} := 18,$

$$\Omega_1(f)(x,y,z,z_1) := \begin{pmatrix} (x^3+z)(x^2-y)+z_1 \\ (x^3+y)(x^2-z)+5z_1 \\ (x^3+z)(x^3+y)(z^2-y)+18z_1 \\ x+y+z-1+z_1 \end{pmatrix};$$

- *compute* $T_1 = \{t_1,...,t_{29}\} \subset \mathbb{C}^4$ *the set of solutions of* $\Omega_1(f)(x,y,z,z_1)$ *using the library "solve.lib" in* SINGULAR *;*

- $\widehat{W_1} = \{w_1,...,w_7 \mid \exists t_i \in T_1 : t_i = (w_i,0), i = 1,...,7\} \subset \mathbb{C}^3$ *the 1-witness point super set corresponding to the pure 1-dimensional component of* X*;*

- $S_1 = T_1 \setminus \{(x,y,z,z_1) \in T_1 \mid z_1 = 0\};$

- *compute*

$$T_0 = V(t.\Omega_1(f)(x,y,z,z_1) + (1-t).\begin{pmatrix} \Omega_0(f)(x,y,z) \\ z_1 \end{pmatrix}),$$

  using the homotopy function technique implemented in BERTINI *with the start system* $\Omega_1(f)(x,y,z,z_1)$ *and the start solution set* S_1 *as t goes from 1 to 0.*

- $\widehat{W_0} = T_0 \subset \mathbb{C}^3$ *the 0-witness point super set corresponding to the pure 0-dimensional component of* X*;*

- $\widehat{W} = \{\widehat{W_0}, \widehat{W_1}\}$ *the witness point super set for* X*.*

We note that we did not need to define a start system (with given solutions) to compute the witness point super set.

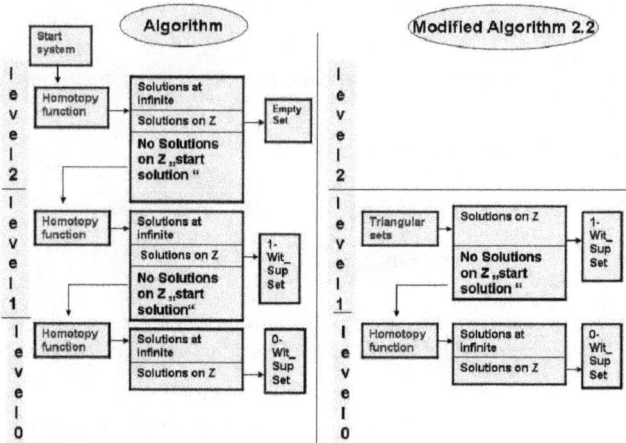

Figure 2.2: Flow Diagram of the Example 2.2.2, the modified algorithm, *Algorithm2*, uses triangular sets in the first level therefore it does not need to define a start system, whose solutions are known. While the algorithm in [26],[33] starts with level two and uses the homotopy continuation method therefore it needs to define a start system, whose solutions are known.

Listing 2 Witness Point Super Set: WitSupSet(F)

Input: $F = \{F_1, ..., F_n\} \subset \mathbb{C}[x_1, ..., x_N]$.

Output: $\{f_1, .., f_N\}$, $\{\widehat{W_r}, .., \widehat{W_d}\}$, L. $\{f_1, .., f_N\}$ a square system, $\widehat{W_i}$ a witness point super set corresponding to a pure i-dimensional component of $V(f_1, ..., f_N)$, L a set of generic linear polynomials.

$f = \{f_1, ..., f_N\}$ reduction of $F = \{F_1, ..., F_n\}$ to a square system (cf. [26], [28],[33]);

$d = dim(V(f_1, ..., f_N))$ (using Gröbner basis cf. [13],[17]);

$r = N - rank(f)$, rank(f) the rank of the Jacobian matrix of the system f at a generic point;

$L = \{l_1, ..., l_d\}$ a set of d generic linear polynomials;

if $d = r$ **then**

    compute $T_d = V(f_1, ..., f_N, l_1, ..., l_d)$ (using a solver based on triangular sets cf. [13],[17]);

    $\widehat{W_d} = \{(x_1, ..., x_N) \mid (x_1, ..., x_N) \in T_d, \ (x_1, ..., x_N) \in V(F)\}$;

    return $\{f_1, ..., f_N\}$, $\{\widehat{W_d}\}$, L ;

else

    for $i = r$ to d **do**

        if $i = 0$ **then**

            $\Omega_i(f)(x) = f$;

        else

$$\Omega_i(f)(x, z_1, ..., z_i) =: \begin{pmatrix} f_1(x) + \sum_{j=1}^{i} \lambda_{1j} z_j \\ \cdot \\ \cdot \\ f_N(x) + \sum_{j=1}^{i} \lambda_{Nj} z_j \\ l_1 + z_1 \\ \cdot \\ \cdot \\ l_i + z_i \end{pmatrix}$$

        $\lambda_{kj} \in \mathbb{C}$ generic, $k = 1, ..., N$, $j = 1, ..., i$;

    for $i = d$ to r **do**

        if $i = d$ **then**

            compute $T_i = V(\Omega_i(f)(x, z_1, ..., z_i))$ (using a solver based on triangular sets cf. [13],[17]);

        else

            compute $T_i = V(\Omega_i(f)(x, z_1, ..., z_i))$ (using a homotopy function with $\Omega_{i+1}(f)(x, z_1, ..., z_i, z_{i+1})$ as start system and S_{i+1} as start solution set cf. [26],[28],[33]);

        $\widehat{W_i} = \{(x_1, ..., x_N) \mid (x_1, ..., x_N, 0, ..., 0) \in T_i, \ (x_1, ..., x_N) \in V(F)\}$;

        $S_i = T_i \setminus \{(x_1, ..., x_N, z_1, ..., z_i) \in T_i \mid z_1 = ... = z_i = 0\}$;

    return $\{f_1, ..., f_N\}$, $\{\widehat{W_r}, ..., \widehat{W_d}\}$, L ;

Remark 2.2.1. *With the notations of the algorithm the following facts prove its correctness and explain our modification:*

1. *The positive dimensional irreducible components of $V(F_1, ..., F_n)$ are the same as the positive dimensional irreducible components of $V(f_1, ..., f_N)$. Isolated points of $V(F_1, ..., F_n)$ are isolated points of $V(f_1, ..., f_N)$. (cf. Theorem 2.1.1)*

2. *It follows from Theorem 1.1.2 that the algebraic variety $V(f_1, ..., f_N)$ has no components of dimension smaller then $r := N - rank(f)$. Therefore the modified algorithm starts at dimension r.*

3. *Since $V(f_1, ..., f_N)$ is of dimension d, the witness point super sets in dimension greater than d are empty. Therefore the modified algorithm can stop at dimension d.*

4. *For $i = 0, 1, ..., d$, it follows from Lemma 2.2.1 above that the sets $\widehat{W}_i$ are witness point super sets for the pure i-dimensional components of $V(F_1, ..., F_n)$.*

5. *In [26],[28],[33] the cascade algorithm is used to compute $\widehat{W}_i$. It starts with $i = N - 1$ to compute the witness point super sets $\widehat{W}_i$. It needs to define a start system $G(x) = 0$ for the homotopy continuation method and to know its solutions. We use a Gröbner basis of the ideal defining $V(F_1, ..., F_n)$ to compute the dimension d of $V(F_1, ..., F_n)$, then use the cascade algorithm which starts with $i = d - 1$.*

Example 2.2.3. *Let X be the algebraic variety defined by the system*

$$f(x, y) = \begin{pmatrix} (x^2 + y^2 - 5)(x - 1) \\ (x^2 + y^2 - 5)(y - 2) \end{pmatrix},$$

which is of dimension 1. Define a generic linear space L_1 by the linear polynomial $l_1 = x + y - 1$.

-

$$\Omega_1(f) = \begin{pmatrix} (x^2 + y^2 - 5)(x - 1) + z \\ (x^2 + y^2 - 5)(y - 1) + 2z \\ x + y - 1 + z \end{pmatrix}.$$

- *compute $T_1 := V(\Omega_1(f)) = \{(2, -1, 0), (-1, 2, 0),$*
  $(0.8524, 0.4048, -0.5572), (1.7513, 2.5027, -3.254),$
  $(-0.8037, -2.6075, 4.4113)\}$ using the library "solve.lib" in the computer algebra system SINGULAR .

- $\widehat{W}_1 := \{(2,-1),(-1,2)\}$ *the witness point super set of 1-dimensional component of* X.

- $S_1 := \{(0.8524, 0.4048, -0.5572), (1.7513, 2.5027, -3.254), (-0.8037, -2.6075, 4.4113)\}$ *the start solution set of the following homotopy function.*

-

$$H_1(x,y,z,t) = t.\Omega_1(f) + (1-t).\begin{pmatrix} \Omega_0(f) \\ z \end{pmatrix} =$$

$$t.\begin{pmatrix} (x^2+y^2-5)(x-1)+z \\ (x^2+y^2-5)(y-1)+2z \\ x+y-1+z \end{pmatrix} + (1-t).\begin{pmatrix} (x^2+y^2-5)(x-1) \\ (x^2+y^2-5)(y-1) \\ z \end{pmatrix}$$

- *compute* $T_0 = \{(1,1,0)\}$ *using the homotopy function in the computer algebra system* BERTINI .

- $\widehat{W}_0 = \{(1,1)\}$ *the witness point super set of 0-dimensional component of* X.

SINGULAR Example :

```
LIB"NumerDecom.lib";
ring ring r=0,(x,y,z),dp;
poly f1=(x^3+z)*(x^2-y);
poly f2=(x^3+y)*(x^2-z);
poly f3=(x^3+z)*(x^3+y)*(z^2-y);
ideal I= f1, f2, f3;
list W=WitSupSet(I);
def A=W[1];
setring A;
L;
   L[1]=6*x+3*y+8*z+11
W(0);
  [1]:                    [2]:
     [1]:                    [1]:
        1                      -1
     [2]:                    [2]:
        1                       1.834387
     [3]:                    [3]:
        1                       1.834387
W(1);
```

```
[1]:                              [2]:
    [1]:                              [1]:
         0                                -1
    [2]:                              [2]:
         0                                 1
    [3]:                              [3]:
        -1.375                            -1
[3]:                              [4]:
    [1]:                              [1]:
        -1                                 0
    [2]:                              [2]:
       -4.33333333333                    -3.66666666666
    [3]:                              [3]:
         1                                 0
[5]:
    [1]:
        1.18016638435673168
    [2]:
       -1.6437271187400354618
    [3]:
       -1.6437271187400354618
[6]:
    [1]:
       (-0.59008319217836584+I*0.7064983903871864045)
    [2]:
       (-0.678136440629982269-I*0.385362758393010766)
    [3]:
       (-0.678136440629982269-I*0.385362758393010766)
[7]:
    [1]:
       (-0.59008319217836584-I*0.7064983903871864045)
    [2]:
       (-0.678136440629982269+I*0.385362758393010766)
    [3]:
       (-0.678136440629982269+I*0.385362758393010766)
```

2.3 Remove Junk Points to Compute Witness point sets

The Algorithm 2 computes the witness point super set $\widehat{W}$ for the algebraic variety $X \subset \mathbb{C}^N$ using the embedding of the generic slicing in a space of dimension $N+d$, where d is the dimension of X. If X_i is a pure i-dimensional component of the algebraic variety $X \subset \mathbb{C}^N$, then a generic $(N-i)$-dimensional linear space L_i will meet X_i in a finite set W_i (cf. Theorem1.1.1) and possible components of dimension greater then i. Particularly, $\widehat{W}_i$ is a union of an i-witness point set $W_i \subset X_i$ and may be some points on components of the dimension $j > i$,

$$\widehat{W}_i = W_i \cup J_i, \tag{2.4}$$

where the finite set $J_i \subset \cup_{j=i+1}^d X_j$ is called junk point set of the component X_i.

   The algorithm in [33] uses only homotopy continuation method to remove the junk point set J_i from $\widehat{W}_i$. In this section we will modify this algorithm by using partially Gröbner bases, triangular sets, local dimension and homotopy continuation method to compute the witness point set W_i as follows.

Proof (The correctness of the Algorithm 3).

1. Since the witness point super set $\widehat{W}_i$ is the union of points on the i-dimensional component and points on components of dimension greater then i, then $\widehat{W}_d$ has no junk points, i.e. $W_d := \widehat{W}_d$. From the definition of the witness point set it follows that $s_d := \sharp W_d$ is the degree of the d-dimensional component of $V(f_1, ..., f_n)$.

2. The witness point super sets are computed numerically, that means $w \in \widehat{W}_i$ is an approximate value of a point v on X.
   Let $Z \subset \mathbb{C}^N \times \mathbb{C}^N$ be the algebraic variety defined by the polynomial system $\{f_1 - t_1, ..., f_N - t_N\}$, $t := (t_1, ..., t_N) \in \mathbb{C}^N$ with $\|t\| \leq 10^{-16}$. Define the map $\varphi : Z \subset \mathbb{C}^N \times \mathbb{C}^N \to \mathbb{C}^N$ by $\varphi(x, t) = t$. Then we have

$$Z_{\varphi(v,0)} = V(f_1, ..., f_N) \subset \mathbb{C}^N \text{ and}$$

$$Z_{\varphi(x, f_1(w), ..., f_N(w))} = V(f_1 - f_1(w), ..., f_N - f_N(w)) \subset \mathbb{C}^N.$$

---

[2]w is the numerical approximate solution of the system $f = \{f_1, ..., f_n\}$, i.e. we consider $f(w) = 0$ numerically.

Listing 3 REMOVE JUNK POINT: WITSET(F)

Input: $\{f_1, .., f_N\} \subset \mathbb{C}[x_1, .., x_N]$, $\{\widehat{W}_r, .., \widehat{W}_d\}$ a list of witness point super sets, $L = \{l_1, .., l_d\}$ a set of generic linear polynomials (Output of Algorithm 1).

Output: $\{f_1, ..., f_N\}$, $\{W_r, .., W_d\}$, $L = \{l_1, ..., l_d\}$. W_i a witness point set corresponding to a pure i-dimensional component of $V(f_1, ..., f_N)$.

$W_d = \widehat{W}_d$, $s_d = \sharp W_d$;
for $i = d - 1$ *to* r **do**
    $W_i = \widehat{W}_i$;
    for each point $w \in W_i$ **do**
        compute $t = dim_w Z$ for $Z = V(f_1 - f_1(w), ..., f_N - f_N(w))$ (using a
            Gröbner basis cf. [13],[17]);
        if $t > i$ **then**
            $W_i = W_i \setminus \{w\}$;
    for each point[2] $w \in W_i$ **do**
        if $i = 0$ **then**
            choose $A \subset \mathbb{C}^{d \times N}$ a generic matrix and a generic $\epsilon \in \mathbb{C}^N$,
                $\|\epsilon\| < 10^{-16}$;
            compute $S = V(\{f_1, ..., f_N, A(x - w)\})$, $T = V(\{f_1, ..., f_N, A(x - w - \epsilon)\})$ (using a solver based on triangular sets cf. [13],[17]);
            if $\sharp S = \sharp T$ **then**
                $W_i = W_i \setminus \{w\}$;
        for $j = i + 1$ *to* d **do**
            choose $A \subset \mathbb{C}^{j \times N}$ a generic matrix;
            if $j = d$ **then**
                compute $S = V(\{f_1, ..., f_N, A(x - w)\})$ (using a solver based on
                    triangular sets cf. [13],[17]);
                if $\sharp S = s_d$ **then**
                    $W_i = W_i \setminus \{w\}$;
            else
                compute $S = V(\{f_1, ..., f_N, A(x - w)\})$ (using a homotopy
                    function with the start system $\{f_1, .., f_N, l_1, .., l_j\}$ and the start
                    solution W_j cf. [33]);
                if $w \in S$ **then**
                    $W_i = W_i \setminus \{w\}$;
return $\{f_1, ..., f_N\}$, $\{W_r, ..., W_d\}$, L ;

It follows (cf. Proposition 1.1.2)

$$t := dim_w V(f_1 - f_1(w), ..., f_N - f_N(w)) \leq dim_v V(f_1, ..., f_N).$$

If $t > i$, then w must be the approximate value of a point v on a component of dimension greater then i. That means that $w \in J_i$.

3. $w \in \widehat{W_i}$ is the numerical approximate solution of the system $f = \{f_1, ..., f_n\}$. Then we consider $f(w) = 0$ numerically with approximation 10^{-16}; $\|f(w)\| \leq 10^{-16}$.

- If $i = 0$, i.e. $w \in \widehat{W_0}$, then an $(N - d)$-dimensional generic linear space $V(A(x - w))$ meets the algebraic variety $V(f_1, ..., f_N)$ in a finite set S. If the $(N - d)$-dimensional generic linear space $V(A(x - w - \epsilon))$ passing through a neighborhood of w meets $V(f_1, ..., f_N)$ in a set T of the same cardinality, then there exists a neighborhood U of w such that $U \cap X \setminus \{w\} \neq \emptyset$. This implies that w is not an isolated point in $V(f_1, ..., f_N)$, i.e. w is on a component of positive dimension. This implies that $w \in J_i$.

- In case of $i > 0$ the test whether w is on a component of dimension $j \in \{d, d - 1, ..., i + 1\}$ is as follows.

  (a) If $j = d$, the degree of the pure d-dimensional component is s_d. The d-dimensional generic linear space $V(A(x - w)^T)$ through w meets $V(f_1, ..., f_N)$ in a finite set S of cardinality greater or equal to s_d. If $\sharp S = s_d$, then w is on the pure d-dimensional component. It implies that $w \in J_i$.

  (b) If $j < d$, we use the homotopy function to remove the junk points (cf. [33]). In particularly S is given as a slicing of $L(t) = t.V(l_1, ..., l_j) + (1 - t).V(A(x - w)^T)$ and $V(f_1, ..., f_N)$ as t goes from 1 to 0. Since $L_j := V(l_1, ..., l_j)$ is a generic linear space, then $L(t)$ is a general point in the Grassmannian $G(N, N - j)$ (cf. Lemma 1.3.1) as t goes from 1 to 0, i. e, $L(t) \in G(N, N - j) \setminus G^*$ where $G^* \subset G(N, N - j)$ is a proper algebraic subset of non-generic slicing planes with respect to X_j. Since the witness point set $W_j := L_j \cap X_j$ is the start solution set of $L(t)$ where X_j is a pure j-dimensional component, then the solution paths $X_j \cap L(t)$ include all nonsingular solutions of $X_j \cap V(A(x - w)^T)$ at their endpoints as t goes from 1 to 0 (cf. Remark 1.3.1). If $w \in S$, then w is on the j-dimensional component. This implies that $w \in J_i$.

$\square$

Example 2.3.1. *Let X be the algebraic variety defined by the system*

$$f(x,y) = \begin{pmatrix} (x^2 + y^2 - 5)(x - 1) \\ (x^2 + y^2 - 5)(y - 1) \end{pmatrix},$$

which is of dimension 1 (cf. Example 2.2.3). The output of the algorithm, Algorithm 2, is $f = \{f_1, f_2\}$, $L = \{l_1\}$ and the witness point super set $\widehat{W} = \{\widehat{W}_0, \widehat{W}_1\}$, where $f_1 = (x^2 + y^2 - 5)(x - 1)$, $f_2 = (x^2 + y^2 - 5)(y - 1)$, $l_1 = x + y - 1$, $\widehat{W}_0 = \{(1,1)\}$, $\widehat{W}_1 = \{(2,-1),(-1,2)\}$. The witness point sets of X are computed as follows.

- $W_1 := \widehat{W}_1 = \{(2,-1),(-1,2)\}$, $s_1 = \sharp W_1 = 2$.

- $i = 0$, $w = (1,1) \in W_0 := \widehat{W}_0 = \{(1,1)\}$

  - $Z := V(f_1 - f_1(w), f_2 - f_2(w))$,
  - $t := dim_w Z = 0$,
  - $t = i$, then $W_0 := W_0$.

- $i = 0$, $w = (1,1) \in W_0 = \{(1,1)\}$

  - $L_1 := V((\begin{array}{cc} 1 & 2 \end{array}) \cdot \begin{pmatrix} x - 1 \\ y - 1 \end{pmatrix}) = V(x + 2y - 3) \subset \mathbb{C}^2$ *generic linear space through w of the dimension 1,*
  - *compute $S = X \cap L_1 = \{(2.2, 0.4), (1,1), (-1,2)\}$, $\sharp S = 3$,*
  - $L_2 := V((\begin{array}{cc} 1 & 2 \end{array}) \cdot \begin{pmatrix} x - 1 - 10^{-8} \\ y - 1 - 10^{-8} \end{pmatrix}) = V(x + 2y - 3(1 + 10^{-8})) \subset$ $\mathbb{C}^2$ *generic linear space passing through a neighborhood of w,*
  - *compute $T = X \cap L_2 = \{(2.2, 0.4000000002), (-0.99999999, 2)\}$, $\sharp T = 2$,*
  - $\sharp T \neq \sharp S$, then $W_0 := \{(1,1)\}$,

- $j = 1$,

  - $L := V((\begin{array}{cc} 1 & 2 \end{array}) \cdot \begin{pmatrix} x - 1 \\ y - 1 \end{pmatrix}) = V(x + 2y - 3) \subset \mathbb{C}^2$ *generic linear space through w of the dimension 1,*
  - *compute $S = X \cap L = \{(2.2, 0.4), (1,1), (-1,2)\}$, $\sharp S = 3$,*
  - $\sharp S \neq s_1$, then $W_0 := \{(1,1)\}$,

- *Return $W = \{W_0, W_1\}$ the witness point set for X, where $W_0 := \{(1,1)\}$, $W_1 := \widehat{W}_1 = \{(2,-1),(-1,2)\}$.*

SINGULAR Example :

```
LIB"NumerDecom.lib";
ring ring r=0,(x,y,z),dp;
poly f1=(x^3+z)*(x^2-y);
poly f2=(x^3+y)*(x^2-z);
poly f3=(x^3+z)*(x^3+y)*(z^2-y);
ideal I= f1, f2, f3;
list J=WitSet(I);
def A=J[1];
setring A;
L;
   L[1]=6*x+3*y+8*z+11
W(0);
   [1]:
      [1]:
         1
      [2]:
         1
      [3]:
         1
W(1);

   [1]:                        [2]:
      [1]:                        [1]:
         0                          -1
      [2]:                        [2]:
         0                          1
      [3]:                        [3]:
         -1.375                     -1
   [3]:                        [4]:
      [1]:                        [1]:
         -1                         0
      [2]:                        [2]:
         -4.33333333333             -3.66666666666
      [3]:                        [3]:
         1                          0
   [5]:
      [1]:
         1.18016638435673168
      [2]:
```

```
            -1.6437271187400354618
    [3]:
            -1.6437271187400354618
 [6]:
    [1]:
            (-0.59008319217836584+I*0.7064983903871864045)
    [2]:
            (-0.678136440629982269-I*0.385362758393010766)
    [3]:
            (-0.678136440629982269-I*0.385362758393010766)
 [7]:
    [1]:
            (-0.59008319217836584-I*0.7064983903871864045)
    [2]:
            (-0.678136440629982269+I*0.385362758393010766)
    [3]:
            (-0.678136440629982269+I*0.385362758393010766)
```

2.4 Partition Witness Point Sets

In this section we show that the monodromy action on an algebraic variety Z and the zero sum relation are sufficient to find the breakup of the k-witness point set W_k into irreducible k-witness point sets. We present here a modified version of the algorithms described in [25],[32].

Let Z be a pure k-dimensional algebraic variety in $\mathbb{C}^N$, and $Z = \cup_{i=1}^r Z_i$ be the irreducible decomposition of Z. Let $\pi : \mathbb{C}^N \longrightarrow \mathbb{C}^k$ be a generic projection and let $l \subset \mathbb{C}^k$ be a general line. Consider

- $\mathbb{W}_l := \pi^{-1}(l) \cap Z$ a set of r different curves in $\mathbb{C}^N$.

- U the non-empty open subset of l consisting of all points $x \in l$ with $\pi^{-1}(x)$ transversal to Z.

- $W := \pi^{-1}(x) \cap Z$ for a generic element $x \in U$, and V a non-empty subset of W.

- $W_i := \pi^{-1}(x) \cap Z_i$ for an irreducible k-dimensional component Z_i of Z.

- $\lambda : \mathbb{C}^N \longrightarrow \mathbb{C}$ a linear function, one-to-one on W.

- For $y \in U$, let V_y be a subset of $\pi^{-1}(y) \cap Z$ defined by

$$V_y := \{z \mid z \text{ on a curve in } \mathbb{W}_l \text{ through a point of } V\}.$$

We define a function $s : U \longrightarrow \mathbb{C}$ by

$$s(y) = \sum_{z \in V_y} \lambda(z),$$

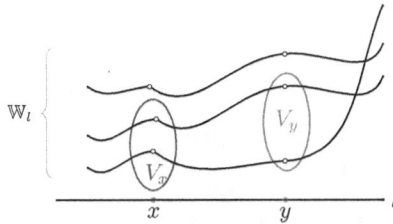

Theorem 2.4.1. *Let l, U, W, V, W_i for $i = 1, ..., r$, and the functions λ, s be as above.*
If the function s is continuous and $V \cap W_i \neq \emptyset$ for some $i \in \{1, ..., r\}$, then $W_i \subseteq V$.

Before proving the theorem we illustrate it by an example.

Example 2.4.1. *Let Z be the curve in $\mathbb{C}^2$ defined by the polynomial $f(x, y) = (x^2 + y^2 - 5)(x - 2y - 3)$. Let L_1 be the line in $\mathbb{C}^2$ defined by the polynomial $l_1 = x + y - 3$. We define a homotopy function :*

$$h(t, x(t), y(t)) := \left(\begin{array}{c} \alpha(t) \\ f(x(t), y(t)) \end{array} \right).$$

$$\alpha(t) = (1 - t)l_0 + tl_1 = x + y - 2t - 1, \text{ where } l_0 = x + y - 1.$$

Then with conditions above $\alpha(t)$ maps a point in $L_1 \cap Z$ to a point in $L_0 \cap Z$ as t goes from 1 to 0, L_0 the line defined by l_0.

Proof (Theorem 2.4.1). Assume that $W_i \nsubseteq V$. Since $W_i \cap V \neq \emptyset$, then there are $a, b \in W_i$ such that a is not in V and b in V. Let $a_1, ..., a_r$ denote the points of the set $V \setminus \{b\}$. By Proposition 1.4.1, there is a loop α in the fundamental group $\Pi_1(U, \pi^{-1}(x))$ with $\alpha(0) = \alpha(1)$ which takes a_j to a_j for all j=1,...,r, and interchanges a and b.

Since α is a continuous loop and $s : U \longrightarrow \mathbb{C}$ is continuous, the composition $s \circ \alpha : [0,1] \longrightarrow \mathbb{C}$ is continuous and

$$s(\alpha(1)) = s(\alpha(0))$$

$$\lambda(a) + \sum_{j=1}^{r} \lambda(a_j) = \lambda(b) + \sum_{j=1}^{r} \lambda(a_j),$$

as t goes from 1 to 0. This implies that $\lambda(a) = \lambda(b)$. But this contradicts the fact that λ is one-to-one on W. Thus W_i is a subset of V. $\qquad\square$

Example 2.4.2. *Let Z be the curve in $\mathbb{C}^2$ defined be the polynomial $f(x,y) = (y-x)(y-2x)(y-3x)$, and $Z = Z_1 \cup Z_2 \cup Z_3$ be the irreducible decomposition. Let $\pi : \mathbb{C}^2 \longrightarrow \mathbb{C}$ be the projection given by $\pi(x,y) = x$, and $\lambda : \mathbb{C}^2 \longrightarrow \mathbb{C}$, $\lambda(x,y) = y$.*
Note that the restriction of π to Z, π_Z is proper and generically three-to-one with degree 3 equal to the degree of Z. λ is one-to-one on the fiber $\pi^{-1}(y) = \{(x,x),(x,2x),(x,3x)\}$. Let L be the line defined by the linear polynomial $l(x,y) = x+y-2$. L intersects Z in the finite set $W := \{(1,1),(\frac{2}{3},\frac{4}{3}),(\frac{1}{2},\frac{3}{2})\}$. Let $V := \{(1,1),(\frac{2}{3},\frac{4}{3})\} \subset W$. The function $\sum_{v \in V} \lambda(v)$ given by $\lambda(x,x) + \lambda(x,2x) = x + 2x = 3x$ is continuous. By the theorem above if an irreducible 1-witness point set W_1 contains $\{(1,1)\}$ or $\{(\frac{2}{3},\frac{4}{3})\}$, then W_1 is a subset of V.

Now we will explain our modification of the algorithm to compute the irreducible witness point sets.
Let $Z_k = \cup_{i=1}^{r} Z_{ki}$ be the union of the irreducible k-dimensional components of the algebraic variety $Z = V(f_1, ..., f_n)$ and L_k be the linear space in $\mathbb{C}^N$ defined by k generic linear polynomials

$$l_j = c_{j0} + c_{j1}x_1 + ... + c_{jN}x_N.$$

for $j = 1, .., k$ and $i = 0, 1, ..., N$, $c_{ij} \in \mathbb{C}$.
We use the generic linear space L_k to define a projection
$\pi : \mathbb{C}^N \longrightarrow \mathbb{C}^{k+1}$, $\pi(x_1, ..., x_N) := (z_1, ..., z_k, z_{k+1})$ as follows:

$$\begin{pmatrix} x_1 \\ \cdot \\ \cdot \\ \cdot \\ \cdot \\ x_N \end{pmatrix} \mapsto \begin{pmatrix} z_1 \\ \cdot \\ \cdot \\ \cdot \\ z_k \\ z_{k+1} \end{pmatrix} := \begin{pmatrix} c_{11} & c_{12} & \cdot & \cdot & c_{1N} \\ c_{21} & c_{22} & \cdot & \cdot & c_{1N} \\ \cdot & \cdot & \cdot & \cdot & \cdot \\ \cdot & \cdot & \cdot & \cdot & \cdot \\ c_{k1} & c_{k2} & \cdot & \cdot & c_{kN} \\ p_1 & p_2 & \cdot & \cdot & p_N \end{pmatrix} \cdot \begin{pmatrix} x_1 \\ \cdot \\ \cdot \\ \cdot \\ \cdot \\ x_N \end{pmatrix},$$

$p_1, ..., p_N \in \mathbb{C}$ randomly chosen.

Set $\lambda(x_1, ..., x_N) := z_{k+1}$ and $l := V(z_1, ..., z_{k-1}) \subset \mathbb{C}^k$ the coordinate axis z_k as in the theorem above. Let $L_{k,y}$ be the linear space defined by the linear polynomials $l_1, ..., l_{k-1}$ and $l_{k,y} := y + c_{k1}x_1 + ... + c_{kN}x_N$. Let $W_y := L_{k,y} \cap Z_k$ be the k-witness point set. For $y = c_{k0}$, we fix a non-empty subset $V = V_y \subset W_y$. In general let V_y be the subset of W_y consisting of all points which are on a curve in $\pi^{-1}(l) \cap Z_k$ through a point of V. To compute V_y we use the homotopy function

$$H(x(t), t) = t. \begin{pmatrix} l_1 \\ . \\ . \\ l_k \\ f_1 \\ . \\ . \\ f_n \end{pmatrix} + (1-t). \begin{pmatrix} l_1 \\ . \\ . \\ l_k - c_{k0} + y \\ f_1 \\ . \\ . \\ f_n \end{pmatrix}$$

as t goes from 1 to 0 using V as start solution set. Note that $\sharp V_y = \sharp V$.
Define the function $s : \mathbb{C} \longrightarrow \mathbb{C}$ by

$$s(y) := \sum_{(x_1, ..., x_N) \in V_y} \lambda(x_1, ..., x_N).$$

To test the linearity of s, we take three values of y in $\mathbb{C}$, say a, b, c.
If there exist $A, B \in \mathbb{C}$ such that

$$(s(a) = Aa + B, s(b) = Ab + B) \implies s(c) = Ac + B, \qquad (2.5)$$

then s is linear.

So far this is the approach which can be found in [25]. We now explain a modification.

The condition (2.5) of the linearity above is equivalent to the following equation

$$s(a)(b - c) + s(b)(c - a) + s(c)(a - b) = 0. \qquad (2.6)$$

If $W_{kj} \cap V_a \neq \emptyset$ for some $j \in \{1, ..., r\}$ and the condition (2.5) is true, then $W_{kj} \subseteq V_a$ (cf. Theorem 2.4.1). Let

$$Z(y) := \{z = \sum_{t=1}^{N} p_t v_t \mid v = (v_1, ..., v_N) \in V_y, p = (p_1, ..., p_N) \in \mathbb{C}^N\}.$$

Then

$$s(y) = \sum_{v \in V_y} \lambda(v) = \sum_{v \in V_y} (\sum_{t=1}^{N} p_t v_t) = \sum_{z \in Z(y)} z.$$

The continuation of the homotopy function implies that the i-th points in the sets V_a, V_b and V_c are on the same irreducible component. Let $V_a := \{v_1, ..., v_m\}$, $V_b := \{\overline{v}_1, ..., \overline{v}_m\}$ and $V_c := \{\hat{v}_1, ..., \hat{v}_m\}$ be the sets computed by using the homotopy function above . Let $Z(a) := \{a_1, ..., a_m\}$, $Z(b) := \{b_1, ..., b_m\}$ and $Z(c) := \{c_1, ..., c_m\}$ be the sets corresponding to the set V_a, V_b and V_c respectively.

From (2.6) we obtain an equivalent condition to (2.5)

$$(b - c) \sum_{i=1}^{m} a_i + (c - a) \sum_{i=1}^{m} b_i + (a - b) \sum_{i=1}^{m} c_i = 0. \tag{2.7}$$

The condition (2.7) is called zero sum relation (cf. [10]) of a given subset $V_a \subseteq W$ denoted by $ZSR(V_a)$. The sets V_a, V_b and V_c have distinct points and the same cardinality m, then obviously

$$ZSR(V_a) = \sum_{a_i \in V_a} ZSR(\{a_i\}). \tag{2.8}$$

where $ZSR(\{a_i\}) = (b - c)a_i + (c - a)b_i + (a - b)c_i$ is defined as the zero sum relation of a given point in V_a.

The following algorithm, Algorithm 4, computes irreducible witness point sets, where its correctness follows from Theorem 2.4.1.

We give an example, Example 2.4.3 below, of a pure 2-dimensional variety Z which is a union of two 2-dimensional irreducible components Z_1 and Z_2. Z_1 is of degree three and Z_2 is of degree two. The 2-witness point set W for Z is given as a finite subset of Z consisting of five points $\{w_1, w_2, w_3, w_4, w_5\}$. Z_1 should contain three points $W_1 := \{w_1, w_2, w_3\}$ and the remaining points $W_2 := \{w_4, w_5\}$ are on Z_2. The algorithms (cf. [25],[32]) use the homotopy function at least nine times to breakup W into W_1 and W_2. We will show below that we do not need more than five times to use the homotopy function to breakup W into W_1 and W_2.

Example 2.4.3. *Let Z be the algebraic variety of dimension two in $\mathbb{C}^3$ defined by the polynomial $f(x, y, z) = (x^3 + z)(x^2 - y)$. Let L be the linear space of dimension one in $\mathbb{C}^3$ defined by the linear equations $l_1 = 4x + 7y + 2z + 6$, $l_2 = 5x + 7y + 3z + 6$. Then $W := L \cap Z = \{w_1, w_2, w_3, w_4, w_5\}$,*

---

[3] the $i - th$ point in R corresponds to the $i - th$ point in W_a;

[4] smallest subset with respect to the cardinality.

Listing 4 IRRWITNESSPOINTSET

Input: $\{f_1, ..., f_N\} \subset \mathbb{C}[x_1, ..., x_N]$ system of polynomials, $\{W_r, ..., W_d\}$ list of witness point sets, $L = \{l_1, ..., l_d\}$ set of generic linear polynomials. Where $W_k = \{w_1, ..., w_{m_k}\}$ are witness point sets for a pure k-dimensional component Z_k of $Z = V(f_1, ..., f_N)$, $k = r, ..., d$ (output of Algorithm 2).

Output: $\{\{W_{r1}, ..., W_{rt_r}\}, ..., \{W_{d1}, ..., W_{dt_d}\}\}$, W_{kr_k} irreducible witness point sets corresponding to a k-dimensional irreducible component Z_{kr_k} of Z_k.

for $k = r$ to d **do**

  a:=c_{k0};

  define L_{ka} to be the linear space defined by the subset $\{l_1, ..., l_k\} \subset L$;

  choose $b, c \in \mathbb{C}$ generic, define L_{kb}, L_{kc} as above;

  $W_a = W_k$, $W_b = \emptyset$, $W_c = \emptyset$, $R = \emptyset$;

  choose $p_1, ..., p_N \in \mathbb{C}$;

  for $i = 1$ to m_k **do**

    compute $\{v_i\} \subset Z \cap L_{k,b}$ and $\{\widehat{v}_i\} \subset Z \cap L_{k,c}$ (using the homotopy function with $\{f_1, ..., f_N, l_1, ..., l_{k-1}, l_{k,a}\}$ as start system and $\{w_i\}$ as start solution);

    compute the zero sum relation[3]of $\{w_i\}$

$$r_i = (a - b)(\sum_{j=1}^{N} p_j \widehat{v}_{ij}) + (b - c)(\sum_{j=1}^{N} p_t w_{ij}) + (c - a)(\sum_{j=1}^{N} p_t v_{ij});$$

    $R = R \cup \{r_i\}$;

  $t_k = 0$;

  while $R \neq \emptyset$ **do**

    if $\sum_{t \in T} t = 0$ and T is a smallest subset[4] of R **then**

    $t_k = t_k + 1$;

    define $W_{kt_k} \subset W_a$ consisting of the points corresponding of the points

      of T;

    $R = R \setminus T$;

  return $\{\{W_{r1}, ..., W_{rt_r}\}, ..., \{W_{d1}, ..., W_{dt_d}\}\}$;

---

where[5] $w_1 = (1, -1.1428571429, -1), w_2 = (0, -0.8571428571, 0),$
$w_3 = (-0.1428571429 + i * 0.9147320339, -0.8163265306 - i * 0.2613520097,$
$0.1428571429 - i * 0.9147320339),$
$w_4 = (-1, -0.5714285714, 1),$
$w_5 = (-0.1428571429 - i * 0.9147320339, -0.8163265306 + i * 0.2613520097,$
$0.1428571429 + i * 0.9147320339).$

We now illustrate the Algorithm 4:

- *Use the linear space L_1 to define the linear projection $\pi : \mathbb{C}^3 \longrightarrow \mathbb{C}^3$ as follows*

$$\pi(x, y, z) := \begin{pmatrix} 4 & 7 & 2 \\ 5 & 7 & 3 \\ 1 & 2 & 3 \end{pmatrix} \begin{pmatrix} x \\ y \\ z \end{pmatrix} = (4x+7y+2z, 5x+7y+3z, x+2y+3z),$$

  where $p_1 = 1, p_2 = 2$ and $p_3 = 3$.

- *Define the linear space $L_{1,c}$ of dimension one in $\mathbb{C}^3$ by the linear equations $l_1 = 4x + 7y + 2z + 6$, $l_c = 5x + 7y + 3z + c$, where c is generically chosen in $\mathbb{C}$. Then*

$$\pi_{Z \cap L_{1,c}}(x, y) = (-6, -c, x + 2y + 3z).$$

- *Define the linear function $\lambda : \mathbb{C}^2 \longrightarrow \mathbb{C}$ by $\lambda(x, y, z) := x + 2y + 3z$.*

- *For $a = 6$, let $V_1 = V_a := \{w_{11} = (1, -1.1428571429, -1)\} \subset W$, $L_{1,a} := L$ the linear space defined by $l_1 = 4x + 7y + 2z + 6$, $l_a = 5x + 7y + 3z + 6$. Then[6] $Z(a) = \{\sum_{v \in V_a} \lambda(v) = w_{11}[1] + 2(w_{11}[2]) + 3(w_{11}[3])\} = \{-4.2857142858\}$.*

- *Let $b = 9$, $L_{1,b}$ the linear space defined by $l_1 = 4x + 7y + 2z + 6$, $l_b = 5x + 7y + 3z + 9$. Compute $V_b := (tL_{1,a} + (1-t)L_{1,b}) \cap Z = \{w_{12} = (1.67169988165715, -0.477628537616333, -4.67169988165716)\}$ as t goes from 1 to 0, using V_a as the start solution. By the using the homotopy method in the computer algebra system BERTINI . Then $Z(b) = \{w_{12}[1] + 2(w_{12}[2]) + 3(w_{12}[3])\} = \{-13.2986568385470012\}$.*

- *Let $c=63$, $L_{1,c}$ the linear space defined by $l_1 = 4x + 7y + 2z + 6$, $l_c = 5x + 7y + 3z + 63$. Compute $V_c := (tL_{1,a} + (1-t)L_{1,c}) \cap Z = \{w_{13} =$*

---

[5]Note that the values of w_i are approximate values. The following equalities are therefore to interpret as approximations of the points w_i.
[6]we use the notation $w_{ij} = (w_{ij}[1], w_{ij}[2], w_{ij}[3])$ for $i = 1, .., 5, j = 1, 2, 3$.

$(3.935100643260828, 14.30425695906836, -60.93510064326094)\}$ *as t goes from 1 to 0, using V_a as the start solution. By the using the homotopy method in the computer algebra system* BERTINI . *Then* $Z(c) = \{w_{13}[1] + 2(w_{13}[2]) + 3(w_{13}[3])\} = \{-150.261687368385272\}.$

$$r_1 := \sum_{a \in Z(a)} (b - c) + \sum_{b \in Z(b)} (c - a) + \sum_{c \in Z(c)} (a - b) =$$
$$= -75.8098062588232524.$$

The zero sum relation set of $V_1 = \{(1, -1.1428571429, -1)\}$ *is*
$R_1 := \{r_1 = -75.8098062588232524\}.$

- *Let* $a = 6$, $V_a := \{w_{11} = (0, -0.8571428571, 0)\} \subset W$, $L_{1,a} := L$ *the linear space defined by* $l_1 = 4x + 7y + 2z + 6$, $l_a = 5x + 7y + 3z + 6$. *Then* $Z(a) = \{\sum_{v \in V_a} \lambda(v) = w_{11}[1] + 2(w_{11}[2]) + 3(w_{11}[3])\} = \{-1.7142857142\}.$

- *Let* $b = 9$, $L_{1,b}$ *the linear space defined by* $l_1 = 4x + 7y + 2z + 6$, $l_b = 5x + 7y + 3z + 9$. *Compute* $V_b := (tL_{1,a} + (1 - t)L_{1,b}) \cap Z = \{w_{12} = (-0.8358499408285809 + i * 1.046869318849985, 0.2388142688081706 - i * 0.2991055196714253, -2.164150059171436 - i * 1.046869318849981)\}$ *as t goes from 1 to 0, using V_a as the start solution.* $Z(b) = \{w_{12}[1] + 2(w_{12}[2]) + 3(w_{12}[3])\} = \{-6.8506715807265477 - i * 2.6919496770428086\}.$

- *Let* $c = 63$, $L_{1,c}$ *the linear space defined by* $l_1 = 4x + 7y + 2z + 6$, $l_c = 5x + 7y + 3z + 63$. *Compute* $V_c := (tL_{1,a} + (1 - t)L_{1,c}) \cap Z = \{w_{13} = (-1.967550321630417 + i * 3.257877039491183, 15.99072866332302 - i * 0.9308220112831772, -55.03244967836969 - i * 3.257877039491242);\}$ *as t goes from 1 to 0, using V_a as the start solution.* $Z(c) = \{w_{13}[1] + 2(w_{13}[2]) + 3(w_{13}[3])\} = \{-135.083442030093447 - i * 8.3773981015488974\}.$

$$r_2 := \sum_{a \in Z(a)} (b - c) + \sum_{b \in Z(b)} (c - a) + \sum_{c \in Z(c)} (a - b) =$$
$$= 107.3334745556671221 - i * 128.308937286793398.$$

The zero sum relation set of $V_2 = \{(0, -0.8571428571, 0)\}$ *is*
$R_2 := \{r_2 = 107.3334745556671221 - i * 128.308937286793398\}.$

- *For the other points* $V_3 = \{w_3\}, V_4 = \{w_4\}$ *and* $V_5 = \{w_5\}$, *we found the zero sum relations* $R_3 := \{r_3 = -9.38237104997583366 + i * 127.0170767088\},$
$R_4 := \{r_4 = -31.5236682999307779 + i * 128.3089372867945956\}$ *and*
$R_5 := \{r_5 = 9.382371038077068 - i * 127.0170767088\}$, *respectively.*

- *The set of zero sum relation for all points of W is $R = \cup_{j=1}^{5} R_j = \{r_1, r_2, r_3, r_4, r_5\}$, where i-th point in W corresponds i-th point in R.*

- *Find the smallest subset T of R with $\sum_{t \in T} t = 0$, which corresponds an irreducible witness point set of W. Then we get $T_1 = \{r_3, r_5\}$, $T_2 = \{r_1, r_2, r_4\}$ corresponding to the irreducible witness point sets $W_1 = \{w_3, w_5\}$, $W_2 = \{w_1, w_2, w_4\}$ respectively.*

Remark 2.4.1. *The points of a witness point set are computed approximately by using the homotopy continuation method. Therefore the result of the zero sum relation is only almost zero.*

2.5 Algorithm of the Numerical Irreducible Decomposition

Combining the algorithms 1, 2, 3 and 4, we obtain the complete algorithm, Algorithm 5, to compute the numerical irreducible decomposition of an algebraic variety defined by a polynomial system as follows.

Listing 5 NUMERICAL IRREDUCIBLE DECOMPOSITION: NUMIRRDE-COM(F)

Input: $F = \{F_1, ..., F_n\} \subset \mathbb{C}[x_1, ..., x_N]$ system of n polynomials in N variables.

Output: $(f, \mathbb{W}, L)$, where $f = \{f_1, ..., f_N\} \subset \mathbb{C}[x_1, ..., x_N]$ square system of N polynomial in N variables, $\mathbb{W}$ the list of irreducible witness point sets ordered by the dimension of the components of $V(F)$ and L the list of generic linear polynomials.

$f := re2squ(F)$ reduction to a system of N polynomials and N variables;
$(f, \widehat{W}, L) := WitSupSet(f)$, $\widehat{W}$ list of witness point super sets for $V(F)$, L list of generic linear polynomials;
$(f, W, L) := WitSet(f, \widehat{W}, L)$, W list of witness point sets for $V(F)$;
$(f, \mathbb{W}, L) := IrrWitnessPointSet(f, W, L)$;
return $(f, \mathbb{W}, L)$;

SINGULAR Example:

```
LIB"NumerDecom.lib";
ring ring r=0,(x,y,z),dp;
```

```
poly f1=(x^3+z)*(x^2-y);
poly f2=(x^3+y)*(x^2-z);
poly f3=(x^3+z)*(x^3+y)*(z^2-y);
ideal I= f1, f2, f3;
list W=NumIrrDecom(I);
============================================
============================================
Dimension
0
Number of Components
1
============================================
============================================
Dimension
1
Number of Components
5
The generic Linear Space L
L[1]=6*x+3*y+8*z+11
 def A(1)=W[1];
 setring A(1);  \\ corresponded to 0-dimensional components
 w(1);
     [1]:
         1
     [2]:
         1
     [3]:
         1
 def A(2)=W[2];
 setring A(2);   \\ corresponded to 1-dimensional components
 w(1);                  w(2);
     [1]:                  [1]:
         0                    -1
     [2]:                  [2]:
         0                     1
     [3]:                  [3]:
        -1.375                -1
 w(3);                  w(4);
     [1]:                  [1]:
        -1                     0
     [2]:                  [2]:
```

```
            -4.33333333333              -3.66666666666
       [3]:                       [3]:
        1                          0
 w(5);
     [1]:
        [1]:
           1.18016638435673168
        [2]:
           -1.6437271187400354618
        [3]:
           -1.6437271187400354618
     [2]:
        [1]:
           (-0.59008319217836584+I*0.7064983903871864045)
        [2]:
           (-0.678136440629982269-I*0.385362758393010766)
        [3]:
           (-0.678136440629982269-I*0.385362758393010766)
     [3]:
        [1]:
           (-0.59008319217836584-I*0.7064983903871864045)
        [2]:
           (-0.678136440629982269+I*0.385362758393010766)
        [3]:
           (-0.678136440629982269+I*0.385362758393010766)
```

2.6 Examples and timings with SINGULAR and BERTINI

In this section we provide examples with timings of the algorithm, Algorithm 5 Numerical Irreducible Decomposition, implemented in SINGULAR to compute[7] the numerical irreducible decomposition of a given algebraic variety defined by a polynomial system and compare them with the results of BERTINI .

We tested to versions of the implementations in BERTINI using the cascade algorithm and using the regenerative cascade algorithm. Timings are conducted by using the 32-bit version of SINGULAR 3-1-1 (cf. [13]) and

---

[7]The SINGULAR implementation uses BERTINI to compute the solutions of the homotopy function.

BERTINI 1.2 (cf. [4]) on an Intel® Core(TM)2 Duo CPU P8400 @ 2.26 GHz 2.27 GHz, 4 GB RAM under the Kubuntu Linux operating system.

Let Z be the algebraic variety defined by the following polynomial system:

Example 2.6.1. *(cf. [28]).*

$$f(x,y,z) = \begin{pmatrix} (y-x^2)(x^2+y^2+z^2-1)(x-\frac{1}{2}) \\ (z-x^3)(x^2+y^2+z^2-1)(y-\frac{1}{2}) \\ (y-x^2)(z-x^3)(x^2+y^2+z^2-1)(z-\frac{1}{2}) \end{pmatrix}$$

Example 2.6.2. *(cf. [33],Example 13.6.4).*

$$f(x,y,z) = \begin{pmatrix} x(y^2-x^3)(x-1) \\ x(y^2-x^3)(y-2)(3x+y) \end{pmatrix}$$

Example 2.6.3.

$$f(x,y,z) = \begin{pmatrix} (x^3+z)(x^2-y) \\ (x^3+y)(x^2-z) \\ (x^3+z)(x^3+y)(z^2-y) \end{pmatrix}$$

Example 2.6.4.

$$f(x,y,z) = \begin{pmatrix} x(y^2-x^3)(x-1) \\ x(3x+y)(y^2-x^3)(y-2) \\ x(y^2-x^3)(x^2-y) \end{pmatrix}$$

Example 2.6.5.

$$f(x,y,z) = \begin{pmatrix} (x-1)((x^3+z)+(x^2-y)) \\ (x^3+z)(x^2-y) \\ (x^3+z)(x^2-1) \end{pmatrix}$$

Example 2.6.6.

$$f(x,y,z) = \begin{pmatrix} (y-x^2)(x^2+y^2+z^2-1)(x-\frac{1}{2})+x^5 \\ (z-x^3)(x^2+y^2+z^2-1)(y-\frac{1}{2})+y^4 \\ (y-x^2)(z-x^3)(x^2+y^2+z^2-1)(z-\frac{1}{2})+z^6 \end{pmatrix}$$

Example 2.6.7.

$$f(x_1,x_2,x_3,x_4,x_5) = \begin{pmatrix} x_5^2+x_1+x_2+x_3+x_4-x_5-4 \\ x_4^2+x_1+x_2+x_3-x_4+x_5-4 \\ x_3^2+x_1+x_2-x_3+x_4+x_5-4 \\ x_2^2+x_1-x_2+x_3+x_4+x_5-4 \\ x_1^2-x_1+x_2+x_3+x_4+x_5-4 \end{pmatrix}$$

Example 2.6.8.

$$f(a,b,c,d,e,f,g) = \begin{pmatrix} a^2 + 2de + 2cf + 2bg + a \\ 2ab + e^2 + 2df + 2cg + b \\ b^2 + 2ac + 2ef + 2dg + c \\ 2bc + 2ad + f^2 + 2eg + d \\ c^2 + 2bd + 2ae + 2fg + e \\ 2cd + 2be + 2af + g^2 + f \\ d^2 + 2ce + 2bf + 2ag + g \end{pmatrix}$$

Example 2.6.9.

$$f(x,y) = \begin{pmatrix} -3568891411860300072x^5 + 1948764938x^4 + \\ 3568891411860300072x^2y^2 - 1948764938xy^2 \\ \\ -5105200242937540320x^5y - 1701733414312513440x^4y^2 \\ +11692589628x^5 + 3897529876x^4y + \\ 5105200242937540320x^2y^3 + 1701733414312513440xy^4 - \\ 11692589628x^2y^2 - 3897529876xy^3 \end{pmatrix}$$

Example 2.6.10.

$$f(x,y,z) = \begin{pmatrix} -356737285367005125x^5 - 92300457164036000x^3y + \\ 1121648050080163317x^2z + 290209720279281056yz \\ \\ -356737285367005125x^5 + 887060318883271500x^3z + \\ 1121648050080163317x^2y - 2789081819567309964yz \\ \\ -356737285367005125x^5z^2 + 356737285367005125x^5y + \\ 887060318883271500x^3z^3 - 887060318883271500x^3yz + \\ 1121648050080163317x^2z^3 - 1121648050080163317x^2yz - \\ 2789081819567309964z^4 + 2789081819567309964yz^2 \end{pmatrix}$$

Example 2.6.11.

$$f(x,y,z) = \begin{pmatrix} \begin{aligned} & x^5y^2 + 2x^3y^4 + xy^6 + 2x^3y^2z^2 + 2xy^4z^2 + xy^2z^4 - x^4y^2 \\ & -2x^2y^4 - y^6 - x^5z - 2x^3y^2z - xy^4z - 2x^2y^2z^2 - 2y^4z^2 - \\ & 2x^3z^3 - 2xy^2z^3 - y^2z^4 - xz^5 - 3x^3y^2 - 3xy^4 + x^4z + \\ & 2x^2y^2z + y^4z - 3xy^2z^2 + 2x^2z^3 + 2y^2z^3 + z^5 + 3x^2y^2 + \\ & 3y^4 + 3x^3z + 3xy^2z + 3y^2z^2 + 3xz^3 + 2xy^2 - 3x^2z - 3y^2z - \\ & 3z^3 - 2y^2 - 2xz + 2z \end{aligned} \\ \\ \begin{aligned} & x^6y + 2x^4y^3 + x^2y^5 + 2x^4yz^2 + 2x^2y^3z^2 + x^2yz^4 - \\ & 5x^6 - 10x^4y^2 - 5x^2y^4 - x^4yz - 2x^2y^3z - y^5z - 10x^4z^2 - \\ & 10x^2y^2z^2 - 2x^2yz^3 - 2y^3z^3 - 5x^2z^4 - yz^5 - 3x^4y - \\ & 3x^2y^3 + 5x^4z + 10x^2y^2z + 5y^4z - 3x^2yz^2 + 10x^2z^3 + \\ & 10y^2z^3 + 5z^5 + 15x^4 + 15x^2y^2 + 3x^2yz + 3y^3z + 15x^2z^2 \\ & +3yz^3 + 2x^2y - 15x^2z - 15y^2z - 15z^3 - 10x^2 - 2yz + 10z \end{aligned} \\ \\ \begin{aligned} & x^6y^2z + 2x^4y^4z + x^2y^6z + 2x^4y^2z^3 + 2x^2y^4z^3 + \\ & x^2y^2z^5 - 7x^6y^2 - 14x^4y^4 - 7x^2y^6 - x^6z^2 - 17x^4y^2z^2 - \\ & 17x^2y^4z^2 - y^6z^2 - 2x^4z^4 - 11x^2y^2z^4 - 2y^4z^4 - x^2z^6 - \\ & y^2z^6 + 7x^6z + 18x^4y^2z + 18x^2y^4z + 7y^6z + 15x^4z^3 + \\ & 27x^2y^2z^3 + 15y^4z^3 + 9x^2z^5 + 9y^2z^5 + z^7 + 21x^4y^2 + \\ & 21x^2y^4 - 4x^4z^2 + 13x^2y^2z^2 - 4y^4z^2 - 11x^2z^4 - 11y^2z^4 - \\ & 7z^6 - 21x^4z - 40x^2y^2z - 21y^4z - 24x^2z^3 - 24y^2z^3 - 3z^5 - \\ & 14x^2y^2 + 19x^2z^2 + 19y^2z^2 + 21z^4 + 14x^2z + 14y^2z + \\ & 2z^3 - 14z^2 \end{aligned} \end{pmatrix}$$

Example 2.6.12.

$$f(x,y,z) = \begin{pmatrix} x(y^2 - x^3)(x-1) + y^2 \\ x(y^2 - x^3)(y-2)(3x+y) + x^3 \end{pmatrix}$$

Example 2.6.13.

$$f(x,y,z) = \begin{pmatrix} (x^3 + z)(x^2 - y) + x^4 \\ (x^3 + y)(x^2 - z) + y^3 \\ (x^3 + z)(x^3 + y)(z^2 - y) + z^5 \end{pmatrix}$$

Example 2.6.14. *cyclic 4-roots problem.(cf.[7],[8]).*

Example 2.6.15. *cyclic 5-roots problem.(cf.[7],[8]).*

Example 2.6.16. *cyclic 6-roots problem.(cf.[7],[8]).*

Example 2.6.17. *cyclic 7-roots problem.(cf.[7],[8]).*

Example 2.6.18. *cyclic 8-roots problem.(cf.[7],[8]).*

Example 2.6.19.

$$f(x_{11}, x_{12}, x_{13}, x_{14}, x_{15}, x_{21}, x_{22}, x_{23}, x_{24}, x_{25}, x_{31}, x_{32}, x_{33}, x_{34}, x_{35}) =$$

$$= \begin{pmatrix} -x_{12}x_{21} + x_{11}x_{22} \\ -x_{13}x_{22} + x_{12}x_{23} \\ -x_{14}x_{23} + x_{13}x_{24} \\ -x_{15}x_{24} + x_{14}x_{25} \\ -x_{22}x_{31} + x_{21}x_{32} \\ -x_{23}x_{32} + x_{22}x_{33} \\ -x_{24}x_{33} + x_{23}x_{34} \\ -x_{25}x_{34} + x_{24}x_{35} \end{pmatrix}$$

Table 2.1 summarizes the results of the timings to compute the numerical decomposition[8].

Remark 2.6.1. *The timings show that for an increasing number of variables the original method of (cf. [25],[26],[28],[32],[33]) becomes more efficient. One reason is that the computation of triangular sets which is used in SIN-GULAR for solving polynomial systems is expensive in this case. Therefore the algorithms, Algorithm 2 and Algorithm 3, become slow in this situation. This is not true for the Algorithm 4.*
Replacing the solving of polynomial system using triangular sets by homotopy function methods but keeping the computation of the dimension and starting in this dimension is more efficient in a case of a large number of variables.

---

[8](re) means using the regenerative cascade algorithm instead of the cascade algorithm

Example	BERTINI	BERTINI (re)	SINGULAR
2.6.1	134.45s	39s	36.07
2.6.2	3.08s	2.5s	1.49s
2.6.3	1min 21.28s	27.4s	4.02s
2.6.4	18.56s	2.7s	1.77s
2.6.5	15.36s	8.6s	1.29s
2.6.6	4min 13s	15min 2s	2min 27s
2.6.7	4.73s	6s	0.37s
2.6.8	5.84s	8s	1s
2.6.9	16s	7s	2s
2.6.10	2min 57s	28s	2min 35s
2.6.11	44min 56s	2min 37s	4min 3s
2.6.12	1.83s	1.6s	0.39s
2.6.13	3min 29s	10min 43s	1.69s
2.6.14	1.43s	4.3s	0.79s
2.6.15	3.54s	10s	0.57s
2.6.16	3min 23.26s	2min 29s	1.43s
2.6.17	2h 11min 57s	32min 17s	stopped after 5h
2.6.18	19h48min 17s	6h45min2s	stopped after 50h
2.6.19	1min 57s	51s	stopped after 3h

Table 2.1: Total running times for the computing a numerical irreducible decomposition of the examples above

Chapter 3

Numerical Primary Decomposition of Algebraic Varieties

The numerical primary decomposition of an algebraic variety is computed in [18] depending on the algorithms in [25],[26],[28],[29],[30],[33],[34]; they compute the numerical irreducible decomposition. Depending on the modified algorithm, Algorithm 5 Numerical Irreducible Decomposition, we will formulate in this chapter an algorithm and give some examples in SINGULAR to compute the numerical primary decomposition.

3.1 Definition of Numerical Primary Decomposition

Definition 3.1.1. *Let I be an ideal in the polynomial ring $R := \mathbb{C}[x_1, ..., x_N]$.*

- *The primary decomposition denoted by "P.D" of I is defined as a finite set $\{J_1, ..., J_r\}$ of primary ideals such that $I = \cap_{i=1}^{r} J_i$. "P.D" is called minimal if $\cap_{j \neq i} J_j \nsubseteq J_i$ and $\sqrt{J_i} \neq \sqrt{J_j}$ for $i \neq j$.*

- *The set $Ass(I) = \{P \mid P \in spec(R), \exists f \in R : P = (I : f)\}$ is defined as a set of associated primes of I. Where $spec(R)$ is the set of all prime ideals of R.*

- *The set $Min(I) = \{P \in Ass(I) \mid \nexists Q \in Ass(I) : Q \subsetneq P\}$ is defined as a set of isolated primes of I.*

- *The set $Emb(I) = Ass(I) \setminus Min(I)$ is defined as a set of embedded primes of I.*

Definition 3.1.2. *Let I be an ideal in the polynomial ring $R := \mathbb{C}[x_1, ..., x_N]$. Let $Ass(I) = \{P_1, ..., P_r\}$ be the set of associated primes of I.*

- *The set $Irr(V(I)) = \{V(P) \mid P \in Ass(I)\}$ is defined as a set of the associated components of $V(I)$.*

- *The set $Min(V(I)) = \{Y \mid Y \in Irr(V(I)) \; \nexists Z \in Irr(V(I)) : Y \subsetneqq Z\}$ is defined as a set of the isolated components of $V(I)$.*

- *The set $Emb(V(I)) = Irr(V(I)) \setminus Min(V(I))$ is defined as a set of the embedded components of $V(I)$.*

- *The numerical primary decomposition of I denoted by "N.P.D" is defined as a finite set $\{W_1, ..., W_r\}$ with the following properties, for $i = 1, ..., r$:*

  1. *$\sharp W_i < \infty$,*

  2. *$W_i \subseteq V(P_i)$,*

  3. *$\sharp W_i = deg(V(P_i))$.*

Remark 3.1.1. *The concept of "N.P.D" of an ideal I describes all components $Y \in Irr(V(I))$ in terms of witness point sets, such a witness set W for a component Y describes Y: one can sample the component, determines its degree.*

Example 3.1.1. *Let I be an ideal defined by the polynomial system $f = \{x^2, xy\}$. The numerical primary decomposition of the ideal I is given as the set $W = \{W_1, W_2\}$. Where $W_1 = \{(0,0)\}, W_2 = \{(0, -2.25)\}$ describing the components $Y_1 := \{(x, y) \in \mathbb{C}^2 \mid x = y = 0\}$ and $Y_2 := \{(x, y) \in \mathbb{C}^2 \mid x = 0\}$ respectively.*

3.2 Deflation of Ideals

Definition 3.2.1. *Let I be an ideal generated by the polynomial system $f = \{f_1, ..., f_n\}$ in the polynomial ring $R := \mathbb{C}[x_1, ..., x_N]$ and d be a positive integer.*

- *The deflation matrix $A^{(d)}(x_1, ..., x_N)$ of order d of a polynomial system $f = \{f_1, ..., f_n\}$ is a matrix with elements in R. The rows of $A^{(d)}(x_1, ..., x_N)$ are indexed by $x^\alpha f_j$, where $|\alpha| < d$ and $j = 1, 2, ..., n$. The columns are indexed by partial differential operators*

$\partial^\beta = \frac{\partial^{|\beta|}}{\partial x_1^{\beta_1}\dots\partial x_N^{\beta_N}}$ where $\beta \neq 0$ and $|\beta| \leq d$.

The element at row $x^\alpha f_j$ and column ∂^β is given as

$$\partial^\beta(x^\alpha f_j) = \frac{\partial^{|\beta|}(x^\alpha f_j)}{\partial x_1^{\beta_1}\dots\partial x_N^{\beta_N}}.$$

- The deflation of I is defined as an ideal $I^{(d)}$ generated by the entries $A^{(d)}(x_1, \dots, x_N).(a_1, \dots, a_{B(N,d)})^T$ and $\{f_1, \dots, f_n\}$ in the polynomial ring $R^{(d)} := \mathbb{C}[x_1, \dots, x_N, a_1, \dots, a_{B(N,d)}]$. Where $B(N,d) := \begin{pmatrix} N+d \\ N \end{pmatrix} - 1$.

Remark 3.2.1.   • $A^{(d)}(x_1, \dots, x_N)$ has N_r rows and N_c columns. Where $N_r := n.\begin{pmatrix} N+d-1 \\ N \end{pmatrix}$ and $N_c := \begin{pmatrix} N+d \\ N \end{pmatrix} - 1$.

- For $d = 1$, the deflation matrix $A^{(1)}(x_1, \dots, x_N)$ is the Jacobian matrix of the polynomial system $f = \{f_1, \dots, f_n\}$.

Proposition 3.2.1. The algebraic variety $X^{(d)} := V(I^{(d)}) \subset \mathbb{C}^{N+B(N,d)}$ is well defined.

Proof. (cf. [18]) □

Example 3.2.1. Let I be an ideal defined by the polynomial system $f = \{x^2, xy\}$ in $R = \mathbb{C}[x,y]$.
The deflation matrix of order $d = 1$ of the system f is the matrix

$$A^{(1)}(x,y) = \begin{pmatrix} 2x & 0 \\ y & x \end{pmatrix}$$

The deflation ideal of order $d = 1$ of I is the ideal $I^{(1)}$ generated by the system $\{x^2, xy, 2xx_1, yx_1 + xx_2\}$ in the polynomial ring $R^{(1)} := \mathbb{C}[x, y, x_1, x_2]$.

The deflation matrix of order $d = 2$ of the system f is

$$A^{(2)}(x,y) = \begin{pmatrix} 2x & 0 & 2 & 0 & 0 \\ y & x & 0 & 1 & 0 \\ 3x^2 & 0 & 6x & 0 & 0 \\ 2xy & x^2 & 2y & 0 & 0 \\ 2xy & x^2 & 2y & 0 & 0 \\ y^2 & 2xy & 0 & 2y & 2x \end{pmatrix}$$

The deflation ideal of order $d = 2$ of I is the ideal $I^{(2)}$ generated by the system $\{x^2, xy, 2xx_1 + 2x_3, yx_1 + xx_2 + x_4, 3x^2x_1 + 6xx_3, 2xyx_1 + x^2x_2 + 2yx_3, y^2x_1 + 2xyx_2 + 2yx_4 + 2xx_5\}$ in the polynomial ring $R^{(2)} := \mathbb{C}[x, y, x_1, x_2, x_3, x_4, x_5]$.

Definition 3.2.2. *Let X be an algebraic variety defined by the ideal I in $\mathbb{C}^N$ and $X^{(d)} \subset \mathbb{C}^{N+B(N,d)}$ be the algebraic variety defined by the deflation of order d of the ideal I. Let $Irr(V(I))$ be the set of associated components of X.*

- *The map $\pi_d : X^{(d)} \to X$ is the natural projection induced by the projection $\pi_d : \mathbb{C}^{N+B(N,d)} \to \mathbb{C}^N$, which maps $(x, a) \mapsto x$.*

- *The deflation of order d of a component $Y \in Irr(V(I))$ is defined as the subset $Y^{(d)} := \overline{\pi_d^{-1}(Y^\circ)}$ of $X^{(d)}$. Where Y° is the subset of generic smooth points that do not belong to other components that do not contain Y.*

- *The component $Y \in Irr(V(I))$ is called visible at order d, if $Y^{(d)}$ is an isolated component of $X^{(d)}$.*

Theorem 3.2.1. *Every component $Y \in Irr(V(I))$ is visible at some order d.*

Proof. (cf. [18])   □

Remark 3.2.2. *Let $I \subset \mathbb{C}[x_1, ..., x_N]$ be an ideal, and Y be a component in $Irr(V(I))$.*

- *The deflation of order d of Y is an irreducible subvariety of X^d.*

- *If Y is visible at order $\hat{d}$ then it is visible at any order $d \geq \hat{d}$. Therefore there is an order d such that all components in $Irr(V(I))$ are visible at d.*

3.3   Algorithm

Depending on the concept of the deflation and the numerical irreducible decomposition using the modified algorithm, Algorithm 5 Numerical Irreducible Decomposition, we formulate an algorithm to compute the numerical primary decomposition of algebraic varieties as follows.

Listing 6 NUMERICAL PRIMARY DECOMPOSITION:$NumPrimDecom(F, d)$

Input: I ideal defined by the polynomial system $F = \{f_1, ..., f_n\} \subset \mathbb{C}[x_1, ..., x_N]$, d positive integer.

Output: $W = \{W_1, ..., W_r\}$ a finite set, where W_i is the witness point set corresponded a component $Y \in Irr(V(I))$.

compute $A^{(d)}(x_1, ..., x_N)$ the deflation matrix of order d of the system f;

compute $B(N, d) := \binom{N + d}{N} - 1$;

let $g_1, ..., g_s$ be the entries of $A^{(d)}(x_1, .., x_N).(a_1, .., a_{B(N,d)})^T$. Define

$$I^{(d)} := < f_1, .., f_n, g_1, ..., g_s >_{\mathbb{C}[x_1, ..., x_N, a_1, .., a_{B(N,d)}]}$$

the deflation ideal of I;

compute $W^{(d)} := \{W_1^{(d)}, ..., W_t^{(d)}\}$ the numerical irreducible decomposition of $I^{(d)}$;

set $W := \emptyset$;

for $i = 1$ *to* t **do**
    compute $W_i = \pi_d(W_i^{(d)})$;
    for $w \in W_i$ **do**
        if $f(w) \neq 0$ **then**
            $W_i := W_i \setminus \{w\}$;
    $W = W \cup \{W_i\}$;
return W ;

Proof. The correctness of the Algorithm 6 follows from Theorem 3.2.1.   □

Example 3.3.1. *Let I be an ideal defined by the polynomial system $f = \{x^2, xy\}$ in $R := \mathbb{C}[x, y]$.*

- $A^{(1)}(x, y) = \begin{pmatrix} 2x & 0 \\ y & x \end{pmatrix}$ *deflation matrix of f;*

- $B(2, 1) = \binom{3}{2} - 1 = 2$;

- *The deflation ideal of order 1 of I is $I^{(1)}$ generated by the system $\{x^2, xy, 2xx_1, yx_1 + xx_2\}$ in $R^{(1)} := \mathbb{C}[x, y, x_1, x_2]$ (cf. Example 3.2.1);*

- *The numerical irreducible decomposition of $I^{(1)}$ is $W^{(1)} = \{W_1^{(1)}, W_2^{(1)}\} \subset \mathbb{C}^4$, where $W_1^{(1)} = \{(0, 0, 0.473, -3.052)\}$, $W_2^{(1)} = \{(0, -2.25, 0, 0.5)\}$;*

- The projection of the components from $\mathbb{C}^4$ to $\mathbb{C}^2$ is given as $\pi_1(W_1^{(1)}) = \{(0,0)\} \subset V(I)$, $\pi_1(W_2^{(1)}) = \{(0,-2.25)\} \subset V(I)$;

- The "N.P.D" of I by the deflation of order 1 is $W = \{\{(0,0)\},\{(0,-2.25)\}\}$.

The numerical primary decomposition "N.P.D" of I by the deflation of order 2:

- $B(2,2) = \begin{pmatrix} 4 \\ 2 \end{pmatrix} - 1 = 5$;

- The deflation ideal of order 1 of I is $I^{(2)}$ generated by the system $\{x^2, xy, 2xx_1 + 2x_3, yx_1 + xx_2 + x_4, 3x^2x_1 + 6xx_3, 2xyx_1 + x^2x_2 + 2yx_3, y^2x_1 + 2xyx_2 + 2yx_4 + 2xx_5\}$ in the polynomial ring $R^{(2)} := \mathbb{C}[x,y,x_1,x_2,x_3,x_4,x_5]$ (cf. Example 3.2.1);

- The numerical irreducible decomposition of $I^{(2)}$ is $W^{(2)} = \{W_1^{(2)}, W_2^{(2)}\} \subset \mathbb{C}^7$, where $W_1^{(2)} = \{(0,1.0264,0,-6.6574,0,0,-2.1885)\}$, $W_2^{(2)} = \{(0,0,-1.1417,-4.1942,0,0,0.5887)\}$;

- The projection of the components from $\mathbb{C}^7$ to $\mathbb{C}^2$ is given as $\pi_1(W_1^{(2)}) = \{(0,-2.25)\} \subset V(I)$, $\pi_1(W_2^{(2)}) = \{(0,0)\} \subset V(I)$;

- The "N.P.D" of I by the deflation of order 2 is $W = \{\{(0,-2.25)\},\{(0,0)\}\}$.

We note that the deflation of the order 1 is sufficient to compute all components.

SINGULAR Example:

```
LIB"NumerDecom.lib";
ring r=0,(x,y,z),dp;
poly f1=z^2;
poly f2=z*(x^2+y);
ideal I=f1,f2;
list W=NumPrimDecom(I,1);  // the order of the deflation is 1
def A=W[1];
setring A;
w(1);
    ==>  [1]:
```

```
                 [1]:
                    3.041019955654102
                 [2]:
                    -3.1321306188268494
                 [3]:
                       0
 w(2);
     ==>
        [1]:
           [1]:
              1.8982822066675845
           [2]:
              -3.6034753361507541
           [3]:
                 0
        [2]:
           [1]:
              -2.3107519248876534
           [2]:
              -5.3395744583719952
           [3]:
                 0
```

Chapter 4

Some Numerical Algebraic Algorithms

We present in this chapter some numerical algorithms, which are based on the concept of the numerical irreducible decomposition of an algebraic variety.

Let $X, Y \subset \mathbb{C}^N$ be algebraic varieties defined by the polynomial systems $f = \{f_1, ..., f_n\}$, $g = \{g_1, ..., g_m\} \subset \mathbb{C}[x_1, ..., x_N]$ respectively. The following algorithms:

1. $Incl(X, Y)$ tests whether X is a subset of Y.

2. $Equal(X, Y)$ tests if X is equal to Y.

3. $Degree(X, i)$ computes the degree of a pure i-dimensional component of X using the definition of the witness point set.

4. Let $\widehat{p}$ be the numerical approximate value of a point p on X. $NumLocalDim(X, \widehat{p})$ computes the local dimension of X at p denoted by $dim_p X$.

In [33] the inclusion test and the computation of the degree depend on the algorithms in [25],[26],[28],[29],[30],[33],[34]; they compute the numerical irreducible decomposition. Depending on the modified algorithm, Algorithm 5 Numerical Irreducible Decomposition, we will present algorithms to test the inclusion of two algebraic varieties and to compute the degree of a pure i-dimensional component.

The algorithm in [28],[32],[33] showed that the local dimension $dim_p X$ could be determined by choosing an appropriate family L_t of linear spaces using the homotopy continuation method.

The other algorithm in [5] used the Dayton-Zeng method (cf. [11]) for computing multiplicity structures to compute numerically the local dimension of the algebraic variety X at the point p.

Depending on the modified algorithm, Algorithm 5 Numerical Irreducible Decomposition, local dimension, Gröbner basis and triangular sets, we will present an algorithm to compute numerically the local dimension $dim_p X$.

4.1   Inclusion

Consider $X, Y \subset \mathbb{C}^N$ reduced algebraic varieties defined by the polynomial systems $f = \{f_1, ..., f_n\}$, $g = \{g_1, ..., g_m\} \subset \mathbb{C}[x_1, ..., x_N]$ respectively.
The algorithm $Incl(X, Y)$ gives a criterion for deciding inclusion of X, Y as follows.

Listing 7 INCLUSION: $Incl(X, Y)$

Input: $f = \{f_1, ..., f_n\}$, $g = \{g_1, ..., g_m\} \subset \mathbb{C}[x_1, ..., x_N]$ polynomial systems defining two algebraic varieties X, Y respectively.
Output: $t = 1$ if $X \subset Y$, otherwise $t = 0$.

    compute $\widehat{W} = \{\widehat{W}_0, ..., \widehat{W}_d\}$ a witness point super set for X (output of the algorithm witness point super set (cf. Algorithm 2));
    for $i = 0$ to d **do**
      if $\widehat{W}_i \neq \emptyset$ **then**
        if $(\exists \widehat{w} \in \widehat{W}_i :$ [1] $\|g(\widehat{w})\| > 10^{-16})$ **then**
          return (0);
    return (1) ;

Proof. The correctness of the algorithm $Incl(X, Y)$:
Since the sets $\widehat{W}_0, ..., \widehat{W}_d$ are sets of generic points on X (cf. Definition 2.2.1), then we can write

$$X \nsubseteq Y \text{ if and only if } (\exists i \in \{0, 1, ..., d\}, \ \exists \widehat{w} \in \widehat{W}_i : g(\widehat{w}) \neq 0).$$

$\square$

---

[1]Note that the point $\widehat{w} \in \widehat{W}_i$ is a numerical approximate value of a point v on X, where $\|f(\widehat{w})\| \leq 10^{-16}$.

4.2   Equality

The inclusion test leads to an equality testing algorithm of two algebraic
varieties X, Y, as follows.

Listing 8 EQUALITY: *Equal*(X, Y)

Input: $f = \{f_1, ..., f_n\}$, $g = \{g_1, ..., g_m\} \subset \mathbb{C}[x_1, ..., x_N]$ polynomial systems
  defining two algebraic varieties X, Y respectively.
Output: $t = 1$ if $X = Y$, otherwise return $t = 0$.

  compute $t_1 = Incl(X, Y)$, $t_2 = Incl(Y, X)$;
  return $(t_1.t_2)$;

SINGULAR Example :

```
LIB"NumerAlg.lib";
ring ring r=0,(x,y,z),dp;
poly f1=(x^2+y^2+z^2-6)*(x-y)*(x-1);
poly f2=(x^2+y^2+z^2-6)*(x-z)*(y-2);
poly f3=(x^2+y^2+z^2-6)*(x-y)*(x-z)*(z-3);
ideal I= f1, f2, f3;
poly g1=(x^2+y^2+z^2-6)*(x-1);
poly g2=(x^2+y^2+z^2-6)*(y-2);
poly g3=(x^2+y^2+z^2-6)*(z-3);
ideal J= g1, g2, g3;
def T=Incl(I,J);
     ==>
          Inclusion:
                    1
def T=Incl(J,I);
     ==>
          Inclusion:
                    0
def T=Equal(J,I);
     ==>
          Equality:
                    0
def T=Equal(J,J);
     ==>
          Equality:
                    1
```

4.3 Degree of a pure i-dimensional component of an algebraic variety

The degree of a pure i-dimensional component X_i of an algebraic variety $X \subset \mathbb{C}^N$ is defined to be the number of points of a generic slicing of a linear space $L_i \subset \mathbb{C}^N$ of dimension $N - i$ with X_i.

Listing 9 DEGREE OF A PURE i-DIMENSIONAL COMPONENT: $Deg(X, i)$

Input: $f = \{f_1, ..., f_n\} \subset \mathbb{C}[x_1, ..., x_N]$ a polynomial system defining the algebraic variety X, i a positive integer.
Output: d_i degree of the pure i-dimensional component of X.

  compute $W = \{W_0, ..., W_d\}$ a witness point set for X (output of the
    algorithm witness point set (cf. Algorithm 3));
  if $W_i = \emptyset$ **then**
    $d_i :=$ [2] -1;
  else
    $d_i := \sharp W_i$;
  return d_i ;

Remark 4.3.1. *The correctness of the algorithm $Degree(X, i)$ follows from the definition of the witness point set (cf. Definition 2.0.4). The witness point sets W_i of a pure i-dimensional component X_i of an algebraic variety X are computed as a generic slicing of a generic linear space $L_i \subset \mathbb{C}^N$ of dimension $N - i$ with X_i.*

SINGULAR Example :

```
LIB"NumerAlg.lib";
ring ring r=0,(x,y,z),dp;
poly f1=(x^2+y^2+z^2-6)*(x-y)*(x-1);
poly f2=(x^2+y^2+z^2-6)*(x-z)*(y-2);
poly f3=(x^2+y^2+z^2-6)*(x-y)*(x-z)*(z-3);
ideal I= f1, f2, f3;
def T=Degree(I,1);
     ==>
        The Degree of Component:
        3
def T=Degree(I,2);
```

---

[2]If W_i is empty set then the algebraic variety X has no i-dimensional component.

```
==>
    The Degree of Component:
    2
```

4.4 Local Dimension Computed Numerically

Let $X \subset \mathbb{C}^N$ be an algebraic variety defined by the polynomial system $f = \{f_1, ..., f_n\} \subset \mathbb{C}[x_1, ..., x_N]$. Let $\widehat{p}$ be the numerical approximate value of a point p on X. The maximum dimension of irreducible components of X containing p is called the local dimension of X at p, denoted by $dim_p X$.

The following algorithm, Algorithm 10, computes the local dimension of an algebraic variety X defined by the polynomial system $f = \{f_1, ..., f_n\} \subset \mathbb{C}[x_1, ..., x_N]$ at a point p. Where $\|\widehat{p} - p\|$ is small.

Proof (The correctness of the algorithm $NumLocalDim(X, \widehat{p})$).

- If $f(\widehat{p}) = 0$, then we can use Gröbner basis to compute $t := dim_{\widehat{p}} X$ the local dimension of X at $\widehat{p}$ considering numerically that $\widehat{p}$ and p on the same component of X.

- Since $d := dim(X)$ is the dimension of X, then $dim_p X \leq d$.

  - The local dimension of $\widehat{X}$ at $\widehat{p}$ is less or equal then the local dimension of X at p, where $\widehat{p}$ is the numerical approximate value of a point p on X. If $\widehat{t} = d$, then $t = d$. Since $\widehat{t} := dim_{\widehat{p}} \widehat{X} \leq t := dim_p X \leq d$ (cf. item 2 in the proof of the Algorithm 3).

  - Since the genericity of the matrices $A, \subset \mathbb{C}^{d \times N}$, $B \subset \mathbb{C}^{1 \times d}$, then the d-dimensional linear space $V(A.(x)^T + B^T)$ meets only the pure d-dimensional component X_d of X in a finite set S of the cardinality equal to the degree of X_d, and the d-dimensional linear space $V(A.(x - \widehat{p})^T)$ meets X in a finite set T (cf. Theorem 1.1.1). If $\sharp T = \sharp S$, then $V(A.(x - \widehat{p})^T)$ meets only X_d. That implies p is on X_d, i.e. $dim_p X = d$.

  - The points of the witness point set W_0 are numerical approximate values of the points of the 0-dimensional component X_0 of X.

---

[3] If $\widehat{p}$ is the numerical approximation solution of the system $f = \{f_1, ..., f_n\}$, then we consider $p = \widehat{p}$ and $f(\widehat{p}) = 0$ numerically.

[4] $\widehat{w}$ is a numerical approximate value of an isolated point in X. Then $\widehat{p}$ is near a point p on X, if $\|\widehat{p} - \widehat{w}\| \leq 10^{-16}$, i.e. numerically $\widehat{p} = \widehat{w}$.

Listing 10 NUMERICAL LOCAL DIMENSION: $NumLocalDim(X, \widehat{p})$

Input: $f = \{f_1, ..., f_n\} \subset \mathbb{C}[x_1, ..., x_N]$ polynomial system defining algebraic variety X, $\widehat{p}$ numerical approximate value of a point p on X.

Output: $t := dim_p X$ local dimension of X at p.

    if $f(\widehat{p}) = 0$ **then**

        compute $t := dim_{\widehat{p}} X$ (using Gröbner basis);

    else

        compute $d := dim(X)$ dimension of X (using Gröbner basis);

        compute $\widehat{t} := dim_{\widehat{p}} \widehat{X}$ the local dimension of $\widehat{X}$ at $\widehat{p}$, where $\widehat{X} := V(f_1 - f_1(\widehat{p}), ..., f_n - f_n(\widehat{p}))$ (using Gröbner basis);

        if $\widehat{t} = d$ **then**

            $t := \widehat{t}$;

        else

            compute[3]$S := V(f_1, ..., f_n, A.(x)^T + B^T)$, $T := V(f_1 - f_1(\widehat{p}), ..., f_n - f_n(\widehat{p}), A.(x - \widehat{p})^T)$ (using a solver based on triangular sets, where $A \subset \mathbb{C}^{d \times N}, B \subset \mathbb{C}^{1 \times d}$ generic matrices);

            if $\sharp S = \sharp T$ **then**

                $t := d$;

            else

                compute $W = \{W_0, ..., W_d\}$ a witness point set for X, $L = \{l_1, ..., l_d\}$ a set of generic linear polynomials (output of Algorithm 3);

                if $\exists \widehat{w} \in W_0 : \widehat{w} \, near^4 \, \widehat{p}$ **then**

                    $t := 0$;

                else

                  for $i = d - 1 \, down \, to \, \widehat{t}$ **do**

                    if $W_i \neq \emptyset$ **then**

                      compute $S := V(f_1, ..., f_n, A(x - \widehat{p})^T)$ (using homotopy function with the start solution W_i and the start system $\{f_1, ..., f_n, l_1, ..., l_i\}$);

                      if $\exists s \in S : s \, near \, \widehat{p}$ **then**

                          $t := i$;

                          $i := d + 1$;

    return t ;

Numerically we consider $W_0 = X_0$. Then $dim_p X = 0$ if and only if W_0 contains a numerical approximate value of $\widehat{p}$, i.e, if there exists a point $\widehat{w} \in W_0$ such that $\|\widehat{w} - \widehat{p}\|$ is small.

- For $\widehat{t} \leq i < d$: In particular, S is the generic slicing $L(t) := tq_1 + (1 - t)q_0$ (cf. Lemma 1.3.1) with $V(f)$ as t goes from 1 to 0, where $q_1 \in V(l_1, ..., l_i)$ and $q_0 \in V(A(x - \widehat{p})^T)$. Since the witness point set $W_i := X_i \cap V(l_1, ..., l_i)$ of the pure i-dimensional component X_i is the start solution set of the homotopy function

$$H(x(t), t) := t \cdot \begin{pmatrix} f \\ l_1 \\ \cdot \\ \cdot \\ l_i \end{pmatrix} + (1 - t) \cdot \begin{pmatrix} f \\ A(x - \widehat{p})^T \end{pmatrix},$$

then the solution paths of $H(x(t), t) = 0$ include the set $S = X_i \cap A(x - \widehat{p})^T$ at their endpoints as t goes from 1 to 0 (cf. Remark 1.3.1). If $\exists s \in S : \|s - \widehat{p}\| \leq 10^{-16}$ small, then p is on X_i, i. e, the local dimension of X at p is i. $\qquad\square$

We now illustrate the Algorithm 10 by an example.

Example 4.4.1. *Let $X \subset \mathbb{C}^3$ be an algebraic variety of dimension 2 defined by the polynomial system*

$$f(x, y, z) = \begin{pmatrix} (x^2 + y^2 + z^2 - 6)(x - y)(x - 1) \\ (x^2 + y^2 + z^2 - 6)(x - z)(y - 2) \\ (x^2 + y^2 + z^2 - 6)(x - y)(x - z)(z - 3) \end{pmatrix}.$$

Let $\widehat{p}_1 = (-1.99999, 1, 1)$, $\widehat{p}_2 = (1.99999, 2, 2)$, $\widehat{p}_3 = (1, 2, 2.9999999) \in \mathbb{C}^3$ be the numerical approximate values of points p_1, p_2, p_3 on X, respectively.

The local dimension of X at p_1 is computed as follows.

- *Let*
$$A = \begin{pmatrix} 1 & 2 & 3 \\ 4 & 5 & 6 \end{pmatrix} \subset \mathbb{C}^{2 \times 3}, \ B = \begin{pmatrix} 1 & 2 \end{pmatrix} \subset \mathbb{C}^{1 \times 2}$$

  be the generic matrices.

- *compute* $S := V\left(f, \begin{pmatrix} 1 & 2 & 3 \\ 4 & 5 & 6 \end{pmatrix} \cdot \begin{pmatrix} x \\ y \\ z \end{pmatrix}^T + \begin{pmatrix} 1 & 2 \end{pmatrix}^T\right) =$

  $V(f, x + 2y + 3z + 1, 4x + 5y + 6z + 2) = \{(-0.93669, 1.8733, -1.270031), (1.047809, -2.0956, 0.714476)\};$

- compute[5] $T := V(f, \begin{pmatrix} 1 & 2 & 3 \\ 4 & 5 & 6 \end{pmatrix} \cdot \begin{pmatrix} x + 1.99999 \\ y - 1 \\ z - 1 \end{pmatrix}) =$

  $V(f, x + 2y + 3z - \frac{300001}{100000}, 4x + 5y + 6z - \frac{75001}{25000}) =$
  $\{(-1.99999, 1.00001, 0.99999), (-0.99998, -1.000006, 2.000003)\};$

- $\sharp S = \sharp T = 2.$ This implies that the numerical local dimension of X at p_1 is $\dim_{p_1} X = 2.$

The local dimension of X at p_2 :

- Let

$$A = \begin{pmatrix} 1 & 2 & 3 \\ 4 & 5 & 6 \end{pmatrix} \subset \mathbb{C}^{2 \times 3}, \ B = \begin{pmatrix} 1 & 2 \end{pmatrix} \subset \mathbb{C}^{1 \times 2}$$

  be the generic matrices.

- compute $S := V(f, \begin{pmatrix} 1 & 2 & 3 \\ 4 & 5 & 6 \end{pmatrix} \cdot \begin{pmatrix} x \\ y \\ z \end{pmatrix}^T + \begin{pmatrix} 1 & 2 \end{pmatrix}^T) = V(f, x + 2y + 3z +$

  $1, 4x + 5y + 6z + 2) = \{(-0.93669, 1.8733, -1.270031), (1.047809, -2.0956, 0.714476)\}$

- compute $T := V(f, \begin{pmatrix} 1 & 2 & 3 \\ 4 & 5 & 6 \end{pmatrix} \cdot \begin{pmatrix} x - 1.99999 \\ y - 2 \\ z - 2 \end{pmatrix}) =$

  $V(f, x + 2y + 3z - \frac{1199999}{100000}, 4x + 5y + 6z - \frac{749999}{25000}) =$
  $\{(1.000001, 3.999998, 1.000001), (1.999996, 2.000007, 1.999996),$
  $(2.000001 + i*1.000001, 1.999997 - i*2.000002, 2.000001 + i*1.000001), (2.000001 -$
  $i*1.000001, 1.999997 + i*2.000002, 2.000001 - i*1.000001)\};$

- $\sharp S \neq \sharp T.$ This implies that the point $\widehat{p_2}$ is not near a point on a pure 2-dimensional component of X.

- compute the witness point set $W = \{W_0, W_1, W_2\}$ and the set of the generic linear polynomials $L = \{l_1, l_2\}$ using the Algorithm 3, where $W_0 = \{(1,2,3)\}$, $W_1 = \{(2,2,0), (\frac{4}{3}, \frac{4}{3}, \frac{4}{3}), (3,-2,3)\}$, $W_2 = \{(2,1,1),(1,2,1)\}$, $l_1 = x + y + z - 4$, $l_2 = x + y + 2z - 5$;

  - W_0 has no point $\widehat{w}$ near $\widehat{p_2}$, then $\dim_{p_2} X \neq 0$;

  - compute[6] $S := V(f, \begin{pmatrix} 1 & 2 & 3 \\ 4 & 5 & 6 \end{pmatrix} \cdot \begin{pmatrix} x - 1.99999 \\ y - 2 \\ z - 2 \end{pmatrix}) =$

    $V(f, x + 2y + 3z - \frac{1199999}{100000}, 4x + 5y + 6z - \frac{749999}{25000}) = \{(2,2,1.999999)\};$

  - S has a point $s = (2,2,1.999999)$ such that $\|\widehat{p_2} - s\|$ small;

     – then $dim_{p_2} X = 1$.

The local dimension of X at p_3 :

- *we do the same as above to compute S, T, we will find $\sharp S \neq \sharp T$, then p_3 is not on 2-dimensional component of X;*

- *compute the witness point set $W = \{W_0, W_1, W_2\}$ and the set of the generic linear polynomials $L = \{l_1, l_2\}$ using the Algorithm 3, where $W_0 = \{(1, 2, 3)\}$, $W_1 = \{(2, 2, 0), (\frac{4}{3}, \frac{4}{3}, \frac{4}{3}), (3, -2, 3)\}$, $W_2 = \{(2, 1, 1), (1, 2, 1)\}$, $l_1 = x + y + z - 4$, $l_2 = x + y + 2z - 5$;*

    – *W_0 has the point $\widehat{w} = (1, 2, 3)$ such that $\|\widehat{w} - p\|$ is small;*

    – *then $dim_{p_3} X = 0$;*

SINGULAR Example :

```
LIB"NumerAlg.lib";
int e=12;
ring r=(complex,e,I),(x,y,z),dp;
poly f1=(x2+y2+z2-6)*(x-y)*(x-1);
poly f2=(x2+y2+z2-6)*(x-z)*(y-2);
poly f3=(x2+y2+z2-6)*(x-y)*(x-z)*(z-3);
ideal J=f1,f2,f3;
list p1=0.999999999999+I*0.000000000001,2,3+I*0.000000000001;
def D1=NumLocalDim(J,p1,e);
   ==>
     The Local Dimension:
     0
list p2=-1,0.999999999998,2+I*0.000000000001;
def D2=NumLocalDim(J,p2,e);
   ==>
     The Local Dimension:
     2
list p2=5+I,4.999999999999+I,5+I;
def D2=NumLocalDim(J,p2,e);
```

---

[5]T is computed by using the library "solve.lib" (SINGULAR 3-1-1) in real ring with 5 digits precision.

[6]S is computed by using the homotopy function (BERTINI) with start solution W_1 and start system $\{f, l_1 = x + y + z - 4\}$.

```
==>
    The Local Dimension:
    1
```

Remark 4.4.1. *The points p_1, p_2 and p_3 in the* SINGULAR *Example above are numerical approximate values of points q_1, q_2 and q_3 on the algebraic variety $V(J)$ respectively with the approximation 10^{-12}, i.e, $\|p_i - q_i\| \leq 10^{-12}$ for $i = 1, 2, 3$.*

Chapter 5

Libraries

The SINGULAR implementation uses BERTINI (cf. [4]) to compute the solutions of the homotopy functions. Therefore the use of the following libraries requires to install BERTINI .

5.1 Library "NumerDecom.lib"

```
LIBRARY:  NumerDecom.lib     Numerical Decomposition of Ideals
OVERVIEW:
     The library contains procedures to compute numerical irreducible
     decomposition, and numerical primary decomposition of an algebraic
     variety defined by a polynomial system. The use of the library req-
     uires to install Bertini (www.nd.edu/~sommese/bertini/thedownload)
PROCEDURES:
 re2squ(ideal I);                            reduction to square system
 UseBertini(ideal H,string sv);
        use Bertini to compute the solutions of the homotopy function H
 Singular2betini1(list L);
    adopt the list to be a read file in Bertini as a start solution set
 bertini2Singular(string snp, int q);
    adopt the file of solutions of the homotopy function to be a list in
                                                               SINGULAR
 ReJunkUseHomo(ideal I, ideal L, list W, list w);
                            remove junk points using the homotopy function
 JuReTopDim(ideal J,list w,int tt, int d);
             remove junk points that are on top-dimensional component
 JuReZeroDim(ideal J,list w, int d);
               remove junk points from 0-dimensional component
 WitSupSet(ideal I);                         witness point super set
 WitSet(ideal I);                                 witness point set
 NumIrrDecom(ideal I);             numerical irreducible decomposition
 defl(ideal I, int d);                          deflation of ideal I
 NumPrimDecom(ideal I, int d);         numerical primary decomposition
```

```
LIB "solve.lib";
LIB "matrix.lib";

//////////////////////////////////////////////////////////////////////////
//////////////////////////////////////////////////////////////////////////
proc re2squ(ideal I)
"USAGE:    re2squ(ideal I);I ideal in a polynomial ring of N unknowns
RETURN:    J ideal defined by a system of N polynomials
EXAMPLE: example re2squ;shows an example
"
{
 def S=basering;
 int n=nvars(basering);
 ideal J;
 poly p;
 int N=size(I);
 int i,j;
 if(n==N)
 {
  J=I;
 }
 else
 {
  if(N<n)
  {
   for(i=1;i<=n;i++)
   {
    if(i<=N)
    {
     J[i]=I[i];
    }
    else
    {
     J[i]=0;
    }
   }
  }
  else
  {
   for(i=1;i<=n;i++)
   {
    p=0;
    for(j=N-n;j<=N;j++)
    {
     p=p+random(1,101)*I[j];
    }
    J[i]=I[i]+p;
   }
```

```
 }
}
export(J);
setring S;
return(S);
}
example
{ "EXAMPLE:";echo = 2;
   ring r=0,(x,y,z),dp;
   ideal I= x3+y4,z4+yx,xz+3x,x2y+z;
   def D=re2squ(I);
   setring D;
   J;
}
////////////////////////////////////////////////////////////////////
proc Singular2betini1(list L)
"USAGE: Singular2betini1(list L); L a list
RETURN: text file called start that is used as a start solution set
NOTE:   adopting the list L to be as a start solution of the homtopy
        function
EXAMPLE: Singular2betini1;shows an example
"
{
 write("start",string(size(L)));
 int i,j;
 number a,b;
 string s;
 list LLL;
 for(i=1;i<=size(L);i++)
 {
  LLL=L[i];
  for(j=1;j<=size(LLL);j++)
  {
   a=repart(LLL[j]);
   b=impart(LLL[j]);
   s=string(a)+" "+string(b)+";";
   write("start",s);
  }
 }
 return(0);
}
example
{ "EXAMPLE:";echo = 2;
   ring r=(complex,16,I),(x,y,z),dp;
   list L=list(1,2,3),list(4,5,6+I*2);
   def D=Singular2betini1(L);
}
////////////////////////////////////////////////////////////////////
proc UseBertini(ideal H,string sv)
```

```
"USAGE:    UseBertini(ideal H,string sv);
           H ideal, sv string of the variable of ring
RETURN:    text file called input used in Bertini to compute the solutions
           of the homotopy function H that is existed in file input.text
NOTE:      Need to define a start solution of H
EXAMPLE:   example Use;shows an example
"
{
 int ii,j,k, ph;
 ph=size(H);
 string sff,sf;
 link l=":w ./input";
 write(l,"");
 write(l,"CONFIG");
 write(l,"");
 write(l,"USERHOMOTOPY: 1;");
 write(l,"");
 write(l,"END;");
 write(l,"");
 write(l,"INPUT");
 write(l,"");
 for( ii=1;ii<=size(sv);ii++)
 {
  if((sv[ii]=="(")||(sv[ii]==")"))
  {
   sv=sv[1,ii-1]+sv[ii+1,size(sv)];
  }
 }
 write(l,"variable "+sv+";");
  sff="function";
 if(ph!=1)
 {
  for( ii=1;ii<=ph-1;ii++)
  {
   sff=sff+" f"+string(ii)+",";
  }
  sff=sff+"f"+string(ph)+";";
 }
 else
 {
  sff=sff+" f"+string(1)+",";
 }
 write(l,sff);
 write(l,"pathvariable t;");
 write(l,"parameter s;");
 write(l,"constant gamma;");
 write(l,"");
 write(l,"gamma = 0.8 + 1.1*I;");
 write(l,"");
```

```
write(1,"s=t;");
write(1,"");
short=0;
for( ii=1;ii<=ph;ii++)
{
  sf=string(H[ii]);
  k=find(sf,newline);
 for( j=1;j<=size(sf);j++)
 {
  if(sf[j]=="(")
  {
   if(sf[j+2]==")")
   {
    sf[j]=" ";
    sf=sf[1,j-1]+sf[j+1,size(sf)];
    sf[j+1]=" ";
    sf=sf[1,j]+sf[j+2,size(sf)];
   }
  }
 }
 write(1,"f"+string(ii)+"="+sf+";");
}
write(1,"END;");
system(("sh","bertini<./input"));
return(0);
}
example
{ "EXAMPLE:";echo = 2;
   ring r=0,(x,y,z),dp;
   ideal I= x3+y4,z4+yx,xz+3x,x2y+z;
   string sv=varstr(basering);
   def A=UseBertini(I,sv);
}
/////////////////////////////////////////////////////////////////////////
proc bertini2Singular(string snp, int q)
"USAGE:    bertini2Singular(string snp, int q);
          snp string, q=nvars(basering) integer
RETURN:  re list of the solutions of the homotopy function
EXAMPLE: example bertini2Singular;shows an example
"
{
 def S=basering;
 int nn=nvars(basering);
 int n=q;
 execute("ring R=(complex,18,I),("+varstr(S)+"),dp;");
 number r1,r2;
 list re,ru;
 string sss=read(snp);
 sss=sss+"";
```

```
int i,j,k,m,p;
string ss;
ss=sss[1];
i=2;
while(sss[i]!=" ")
{
 ss=ss+sss[i];
 i++;
}
execute("m="+ss+";");
for(i=1;i<=size(sss);i++)
{
 if(sss[i]=="e")
 {
  if(!((sss[i+1]=="+")||(sss[i+1]=="-")))
  {
   ss=sss[i+1,size(sss)];
   sss=sss[1,i];
   sss=sss+"+"+ss;
  }
 }
}
j=1;
j=find(sss,newline,j)+1;
while(sss[j]==newline){j++;}
for(q=1;q<=m;q++)
{
 for(p=1;p<=n;p++)
 {
  k=find(sss,newline,j);
  ss=sss[j,k-j];
  i=find (ss," ");
  execute("r1="+ss[1,i-1]+";");
  execute("r2="+ss[i+1,size(ss)-i+1]+";");
  ru[p]=r1+I*r2;
  j=k+1;
 }
 j=j+1;
 re[size(re)+1]=ru;
}
export(re);
setring S;
return(R);
}
example
{ "EXAMPLE:";echo = 2;
   ring r = 0,(a,b,c),ds;
   int q=nvars(basering);
   def T=bertini2Singular("nonsingular_solutions",q);
```

```
    re;
}
//////////////////////////////////////////////////////////////////////
proc    WitSupSet(ideal I)
"USAGE: WitSupSet(ideal I);I ideal
RETURN: list of Witness point Super Sets W(i) for i=1,...,dim(V(I)),
        L list of generic linear polynomials and N(0) list of a polynomial
        system of the same number of polynomials and unknowns.
NOTE:   if W(i) = x, then V(I) has no components of dimension i
EXAMPLE: example WitSupSet;shows an example
"
{
 def S=basering;
 int n=nvars(basering);
 ideal II=I;
 int dd=dim(std(I));
 if(n==1)
 {
   ERROR("n=1");
 }
 else
 {
  if(dd==0)
  {
   execute("ring R=0,("+varstr(S)+"),dp;");
   int i,j;
   ideal I=imap(S,I);
   list V(dd),W(dd);
   def T(dd+1)=solve(I,"nodisplay");
   setring T(dd+1);
   W(dd)=SOL;
   ideal N(dd)=imap(S,I);
   export(N(dd));
   ideal LL;
   export(LL);
   int c(0);
   c(0)=0;
   export(c(0));
   export(dd);
   list w(1..size(W(dd)));
   for(i=1;i<=size(W(dd));i++)
   {
    w(i)=W(dd)[i];
    export(w(i));
   }
  "=========================================";
  "=========================================";
   "Dimension";
    dd;
```

```
"Number of Components";
size(W(dd));
setring S;
return(T(dd+1));
}
else
{
matrix MJJ=jacob(I);
int rn=rank(MJJ);
I=imap(S,II);
def rs=re2squ(I);
setring rs;
I=J;
if((n-rn)!=dd)
{
 execute("ring R=0,("+varstr(S)+",z(1..dd)),dp;");
 ideal I=imap(rs,I);
 ideal H(0..n),L,LL,L(1..dd),LL(1..dd),h(1..dd),N(0..dd);
 poly p,p(0..n),e;
int i,j,k,kk,q,qq,t,m,d,jj,rii,c(0),ii;
for(i=1;i<=dd;i++)
{
 p=0;
 for(j=1;j<=n;j++)
 {
  p=p+random(1,2*n+7)*var(j);
 }
 if(i<dd)
 {
  LL[i]=random(2*n+7,4*n+1)+p;
 }
 else
 {
  c(0)=random(4*n+1,5*n+13);
  LL[i]=c(0)+p;
 }
}
export(c(0));
p(0)=0;
for(t=1;t<=n;t++)
{
 for(j=1;j<=dd;j++)
 {
  p(j)=p(j-1)+random(1,2*n+10)*var(n+j);
  h(j)[t]=I[t]+p(j);
 }
}
for(q=1;q<=dd;q++)
{
```

```
 for(i=1;i<=q;i++)
 {
  L(q)[i]=LL[i]+var(n+i);
 }
}
for(i=1;i<=dd;i++)
{
 N(i)=h(i),L(i);
}
for(i=1;i<=n;i++)
{
 N(0)[i]=I[i];
}
ideal JJ=N(0);
if(dim(std(N(dd)))!=0)
{
 "Try Again";
}
else
{
 def T=solve(N(dd),100,"nodisplay");
 setring T;
 execute("ring T(dd+1)=(complex,16,I),("+varstr(S)+"),dp;");
 list M,Y;
 list W(dd),V(dd);
 list SOL=imap(T,SOL);
 Y=SOL;
 number rp,ip,rip;
 for( i=1;i<=size(SOL);i++)
 {
  M=Y[i];
  for(j=dd;j>=1;j--)
  {
   rp=repart(M[n+j]);
   ip=impart(M[n+j]);
   rip=rp^2 + ip^2;
   if(rip<0.0000000000000001)
   {
    M=delete(M,n+j);
    Y[i]=M;
   }
  }
 }
 k=1;
 kk=1;
 for( i=1;i<=size(Y);i++)
 {
  if(size(Y[i])==n)
  {
```

```
  W(dd)[k]=Y[i];
  k=k+1;
  }
 else
 {
  V(dd)[kk]=Y[i];
  kk=kk+1;
  }
}
ideal JJ=imap(S,II);
k=1;
number al,ar,ai,ri;
for(j=1;j<=size(W(dd));j++)
{
 ri=0;
 al=0;
 ai=0;
 ar=0;
 for(ii=1;ii<=size(JJ);ii++)
 {
  for(i=1;i<=n;i++)
  {
   JJ[ii]=subst(JJ[ii],var(i),W(dd)[j][i]);
  }
  al=leadcoef(JJ[ii]);
  ar=repart(al);
  ai=impart(al);
  ri=ar^2+ai^2+ri;
 }
 if(ri<=0.000000000000000001)
 {
  W(dd)[k]=W(dd)[j];
  k=k+1;
 }
}
ideal L(dd)=imap(R,L(dd));
export(L(dd));
export(W(dd));
export(V(dd));
string sff,sf,sv;
int nv(dd)=size(V(dd));
int nv(0..dd-1);
if(size(W(dd))<size(Y))
{
 def SB(dd)=Singular2betini1(V(dd));
}
for( q=dd;q>=n-rn+1;q--)
{
 if(nv(q)!=0)
```

```
{
 int w(q-1)=0;
 execute("ring D(q)=(0,s,gamma),("+varstr(S)+",z(1..q)),dp;");
 string nonsin(q),stnonsin(q);
 ideal H(1..q);
 ideal N(q)=imap(R,N(q));
 ideal N(q-1)=imap(R,N(q-1));
 for(j=1;j<=n+q-1;j++)
 {
  H(q)[j]=s*gamma*N(q)[j]+(1-s)*N(q-1)[j];
 }
 H(q)[n+q]=s*gamma*N(q)[n+q]+(1-s)*var(n+q);
 ideal H=H(q);
 export(H(q));
 string sv(q)=varstr(basering);
 sv=sv(q);
 def Q(q)=UseBertini(H,sv);
 system("sh","rm start");
 nonsin(q)=read("nonsingular_solutions");
 if(size(nonsin(q))>=52)
 {
  def T(q)=bertini2Singular("nonsingular_solutions",
                                       nvars(basering));
  setring T(q);
  list C=re;
  list B,X,A,G;
  for(i=1;i<=size(C);i++)
  {
   B=re[i];
   B=delete(B,n+q);
   C[i]=B;
  }
  X=C;
  if(q>=2)
  {
   for(j=q-1;j>=1;j--)
   {
    for(i=1;i<=size(X);i++)
    {
     A[i]=X[i];
     G=A[i];
     G=delete(G,n+j);
     A[i]=G;
    }
    X=A;
   }
  }
  else
  {
```

```
  X=C;
}
list W(q-1),V(q-1);
ideal JJ=imap(S,II);
k=1;
poly tj;
number al,ar,ai,ri;
for(j=1;j<=size(C);j++)
{
 ri=0;
 al=0;
 ai=0;
 ar=0;
 for(i=1;i<=size(JJ);i++)
 {
  tj=JJ[i];
  for(i=1;i<=n;i++)
  {
   tj=subst(tj,var(i),X[j][i]);
  }
  al=leadcoef(tj);
  ar=repart(al);
  ai=impart(al);
  ri=ar^2+ai^2+ri;
 }
 if(ri<=0.000000000000000001)
 {
   W(q-1)[k]=X[j];
   k=k+1;
 }
 else
 {
  nv(q-1)=nv(q-1)+1;
  V(q-1)[nv(q-1)]=C[j];
 }
}
if(nv(q-1)==size(C))
{
 list W(q-1)=var(1);
}
if(q>=2)
{
 if(nv(q-1)!=0)
 {
  def SB(qq-1)=Singular2betini1(V(q-1));
 }
 else
 {
  for(qq=q-1;qq>=1;qq--)
```

```
       {
        execute("ring T(qq)=(complex,16,I),("+varstr(S)+",z(1..qq)),
                                                            dp;");
        list W(qq-1)=var(1);
       }
      q=1;
     }
    }
   }
   else
   {
    for(qq=q;qq>=1;qq--)
    {
     int w(qq-1);
     execute("ring T(qq)=(complex,16,I),("+varstr(S)+",z(1..qq)),
                                                          dp;");
     list W(qq-1)=var(1);
    }
   }
  }
  else
  {
   for(qq=q;qq>=1;qq--)
   {
    execute("ring T(qq)=(complex,16,I),("+varstr(S)+",z(1..qq)),
                                                         dp;");
    list W(qq-1)=var(1);
   }
  }
 }
 execute("ring D=(complex,16,I),("+varstr(S)+"),dp;");
 for(i=0;i<=dd;i++)
 {
  list W(i)=imap(T(i+1),W(i));
  export(W(i));
 }
 ideal L=imap(R,LL);
 export(L);
 ideal N(0)=imap(R,N(0));
 export(N(0));
 setring S;
 return(D);
 }
}
else
{
 execute("ring R=0,("+varstr(S)+"),dp;");
 int i,j,c(0);
 poly p;
```

```
      ideal LL;
      for(i=1;i<=dd;i++)
      {
       p=0;
       for(j=1;j<=n;j++)
       {
        p=p+random(1,100)*var(j);
       }
       if(i<dd)
       {
        LL[i]=random(101,200)+p;
       }
       else
       {
        c(0)=random(201,300);
        LL[i]=c(0)+p;
       }
      }
      ideal I=imap(S,I);
      ideal N(dd)=I,LL;
      def T=solve(N(dd),100,"nodisplay");
      setring T;
      list W(0..dd);
      W(dd)=SOL;
      export(W(dd));
      for(i=0;i<=dd-1;i++)
      {
       W(i)=var(1);
       export(W(i));
      }
      ideal L=imap(R,LL);
      export(L);
      ideal N(0)=imap(S,I);
      export(N(0));
      setring S;
      return(T);
     }
    }
   }
}
example
{ "EXAMPLE:";echo = 2;
   ring r=0,(x,y,z),dp;
   poly f1=(x2+y2+z2-6)*(x-y)*(x-1);
   poly f2=(x2+y2+z2-6)*(x-z)*(y-2);
   poly f3=(x2+y2+z2-6)*(x-y)*(x-z)*(z-3);
   ideal I=f1,f2,f3;
   def W=WitSupSet(I);
   setring W;
```

```
    W(2);
    // witness point super set of a pure 2-dimensional component of V(I)
    W(1);
    // witness point super set of a pure 1-dimensional component of V(I)
    W(0);
    // witness point super set of a pure 0-dimensional component of V(I)
    L;
    // list of generic linear polynomials
}
//////////////////////////////////////////////////////////////////////
proc ReJunkUseHomo(ideal I, ideal L, list W, list w)
"USAGE:  ReJunkUseHomo(ideal I, ideal L, list W, list w); I ideal,
         L list of generic linear polynomials {l_1,...,l_i}, W list of a
         subset of the solution set of the generic slicing V(L) with V(J),
         w list of a point in V(J)
RETURN:  t=1 if w on an i-dimensional component of V(I),
         otherwise t=0. Where i=size(L)
EXAMPLE: example ReJunkUseHomo;shows an example
"
{
 def S=basering;
 int n=nvars(basering);
 int ii,i,in,j,jjj,jj,k,zi,a,kk,kkk;
 string sf,sff,sv;
 i=size(W);
 in=size(w);
 ideal LL;
 jjj=size(L);
 poly pp;
 for(jj=1;jj<=jjj;jj++)
 {
  for(ii=1;ii<=in;ii++)
  {
   pp=random(1,3*n+1)*(var(ii)-w[ii])+pp;
  }
  LL[jj]=pp;
 }
 export(LL);
 execute("ring R=(complex,16,I),("+varstr(S)+",gamma,s),dp;");
 ideal L=imap(S,L);
 ideal LL=imap(S,LL);
 ideal I=imap(S,I);
 list w=imap(S,w);
 zi=size(I);
 ideal H;
 for(a=1;a<=zi;a++)
 {
  H[a]=s*gamma*I[a]+(1-s)*I[a];
 }
```

```
for(kk=1;kk<=jjj;kk++)
{
 H[kk+zi]=s*gamma*L[kk]+(1-s)*LL[kk];
}
list W=imap(S,W);
def SB1=Singular2betini1(W);
sv=varstr(S);
def Q=UseBertini(H,sv);
system("sh","rm start");
string nonsin=read("nonsingular_solutions");
if(size(nonsin)>=52)
{
 def TT=bertini2Singular("nonsingular_solutions",nvars(basering)-2);
 setring TT;
 list w=imap(S,w);
 list C=re;
 list ww,v;
 number rp,ip,rp(1..size(w)),ip(1..size(w)),irp,t;
 for(k=1;k<=size(C);k++)
 {
  ww=re[k];
  for(jj=1;jj<=size(w);jj++)
  {
   rp(jj)=(repart(ww[jj])-repart(w[jj]))^2;
   ip(jj)=(impart(ww[jj])-impart(w[jj]))^2;
   rp=rp+rp(jj);
   ip=ip+ip(jj);
  }
  irp=ip+rp;
  if(irp<=0.0000000000000000000000001)
  {
   t=1.0;
  }
  else
  {
   t=0.0;
  }
 }
}
else
{
execute("ring TT=(complex,16,I),("+varstr(S)+"),dp;");
list w=imap(S,w);
number t=1.0;
}
export(t);
setring S;
return(TT);
}
```

```
example
{ "EXAMPLE:";echo = 2;
    ring r=(complex,16,I),(x,y,z),dp;
    poly f1=(x2+y2+z2-6)*(x-y)*(x-1);
    poly f2=(x2+y2+z2-6)*(x-z)*(y-2);
    poly f3=(x2+y2+z2-6)*(x-y)*(x-z)*(z-3);
    ideal J=f1,f2,f3;
    poly l1=15x+16y+6z+17;
    poly l2=2x+14y+4z+18;
    ideal L=l1,l2;
    list W1=list(0.5372775295412116,-0.7105339291010922,
                              -2.2817700129167831+I*0),
          list(0.09201175741935605,-1.7791717821935455,
                              1.6810953589677311);
    list w=list(2,2,-131666666/10000000);
    def D=ReJunkUseHomo(J,L,W1,w);
    setring D;
    t;
}
////////////////////////////////////////////////////////////////////////
proc    JuReTopDim(ideal J,list w,int tt, int d);
"USAGE:  JuReTopDim(ideal J,list w,int tt, int d); J ideal,
         w list of a point in V(J), tt the degree of d-dimensional
          component of V(J), d dimension of V(J)
RETURN:  t=1 if w on a d-dimensional component of V(I), otherwise t=0.
EXAMPLE: example JuReTopDim;shows an example
"
{
 def S=basering;
 int n=nvars(basering);
 int i,j,k;
 list iw,rw;
 for(k=1;k<=n;k++)
 {
  rw[k]=repart(w[k]);
  iw[k]=impart(w[k]);
 }
 execute("ring R=real,("+varstr(S)+",I),dp;");
 list iw=imap(S,iw);
 list rw=imap(S,rw);
 ideal J=imap(S,J);
 execute("ring RR=0,("+varstr(S)+",I),dp;");
 ideal J=imap(R,J);
 list iw=imap(R,iw);
 list rw=imap(R,rw);
 ideal L;
 poly p;
 for(i=1;i<=d;i++)
 {
```

```
 p=0;
 for(j=1;j<=n;j++)
 {
  p=p+random(1,100)*(var(j)-rw[j]-I*iw[j]);
 }
 L[i]=p;
}
ideal JJ;
for(i=1;i<=size(J);i++)
{
 p=J[i];
 for(j=1;j<=n;j++)
 {
  p=subst(p,var(j),rw[j]+I*iw[j]);
 }
 JJ[i]=p;
}
poly pp;
pp=I^2 +1;
ideal T=L,J,pp;
int di=dim(std(T));
if(di==0)
{
 def T(d)=solve(T,10,"nodisplay");
 setring T(d);
 number t,ie,re,rt;
 int zi=size(SOL);
 list iw=imap(S,iw);
 list rw=imap(S,rw);
 if(zi==2*tt)
 {
  t=1.0/1;
 }
 else
 {
  t=0.0/1;
 }
}
else
{
 execute("ring T(d)=(complex,16,I),("+varstr(S)+"),dp;");
 "Try Again";
 -----
}
export(t);
setring S;
return(T(d));
}
example
```

```
{ "EXAMPLE:";echo = 2;
    ring r=(complex,16,I),(x,y,z),dp;
    poly f1=(x2+y2+z2-6)*(x-y)*(x-1);
    poly f2=(x2+y2+z2-6)*(x-z)*(y-2);
    poly f3=(x2+y2+z2-6)*(x-y)*(x-z)*(z-3);
    ideal J=f1,f2,f3;
    list w=list(0.5372775295412116,-0.7105339291010922,
                                    -2.2817700129167831);
    def D=JuReTopDim(J,w,2,2);
    setring D;
    t;
}
/////////////////////////////////////////////////////////////////
proc JuReZeroDim(ideal J,list w, int d);
"USAGE:   JuReZeroDim(ideal J,list w, int d);J ideal,
         w list of a point in V(J), d dimension of V(J)
RETURN:  t=1 if w on a positive-dimensional component of V(I),
         i.e w is not isolated point in V(J)
EXAMPLE: example JuReZeroDim;shows an example
"
{
 def S=basering;
 int n=nvars(basering);
 int i,j,k;
 list iw,rw;
 for(k=1;k<=n;k++)
 {
  rw[k]=repart(w[k]);
  iw[k]=impart(w[k]);
 }
 execute("ring R=real,("+varstr(S)+",I),dp;");
 list iw=imap(S,iw);
 ideal J=imap(S,J);
 list rw=imap(S,rw);
 execute("ring RR=0,("+varstr(S)+",I),dp;");
 list iw=imap(R,iw);
 ideal J=imap(R,J);
 list rw=imap(R,rw);
 ideal LL;
 poly p;
 for(i=1;i<=d;i++)
 {
  p=0;
  for(j=1;j<=n;j++)
  {
   p=p+random(1,100)*(var(j)-rw[j]-I*iw[j]);
  }
  LL[i]=p;
 }
```

```
poly pp;
pp=I^2 +1;
ideal TT=LL,J,pp;
def TT(d)=solve(TT,16,"nodisplay");
setring TT(d);
int zii=size(SOL);
execute("ring RR1=0,("+varstr(S)+",I),dp;");
list iw=imap(R,iw);
ideal J=imap(R,J);
list rw=imap(R,rw);
ideal L;
poly p;
for(i=1;i<=d;i++)
{
 p=0;
 for(j=1;j<=n;j++)
 {
  p=p+random(1,100)*(var(j)-rw[j]-I*iw[j]-1/100000000000000);
 }
 L[i]=p;
}
poly pp;
pp=I^2 +1;
ideal T=L,J,pp;
int di=dim(std(T));
if(di==0)
{
 def T(d)=solve(T,16,"nodisplay");
 setring T(d);
 number t;
 int zi=size(SOL);
 list iw=imap(S,iw);
 list rw=imap(S,rw);
 if(zi==zii)
 {
  t=1.0/1;
 }
 else
 {
  t=0.0/1;
 }
}
else
{
 execute("ring T(d)=(complex,16,I),("+varstr(S)+"),dp;");
 "Try Again";
 -----
}
export(t);
```

```
 setring S;
 return(T(d));
}
example
{ "EXAMPLE:";echo = 2;
    ring r=(complex,16,I),(x,y,z),dp;
    poly f1=(x2+y2+z2-6)*(x-y)*(x-1);
    poly f2=(x2+y2+z2-6)*(x-z)*(y-2);
    poly f3=(x2+y2+z2-6)*(x-y)*(x-z)*(z-3);
    ideal J=f1,f2,f3;
    list w1=list(0.5372775295412116,-0.7105339291010922,
                                    -2.2817700129167831);
    def D1=JuReZeroDim(J,w1,2);
    setring D1;
    t;
}
/////////////////////////////////////////////////////////////////////
proc    WitSet(ideal I)
"USAGE: WitSet(ideal I); I ideal
RETURN: lists W(0..d) of witness point sets of i-dimensional components
        of V(J) for i=0,...d respectively, where d the dimension of V(J),
           L list of generic linear polynomials
NOTE:     if W(i)=x, then V(J) has no component of dimension i
EXAMPLE: example WitSet;shows an example
"
{
 def S=basering;
 int n=nvars(basering);
 int ii,i,j,b,bb,k,kk,dt;
 def TJ(0)=WitSupSet(I);
 setring TJ(0);
 ideal LL=L;
 int d=size(LL);
 if(d==0)
 {
  setring S;
  return(TJ(0));
 }
 else
 {
  for( i=0;i<=d;i++)
  {
   list Ww(i)=W(i);
   int z(i)=size(W(i));
   export(Ww(i));
  }
  for(i=d-1;i>=0;i--)
  {
   list W(i)=imap(TJ(0),Ww(i));
```

```
if(size(W(i)[1])>1)
{
 for(j=1;j<=z(i);j++)
 {
  execute("ring Rr(j+i)=(complex,106,I),("+varstr(S)+"),ds;");
  list W(i)=imap(TJ(0),Ww(i));
  list w=W(i)[j];
  ideal J=imap(TJ(0),N(0));
  ideal J(j),K(j);
  for(k=1;k<=size(J);k++)
  {
   J(j)[k]=J[k];
   for(kk=1;kk<=n;kk++)
   {
    J(j)[k]=subst(J(j)[k],var(kk),w[kk]);
   }
   K(j)[k]=J[k]-J(j)[k];
  }
  poly p(1..n);
  for(k=1;k<=n;k++)
  {
   p(k)=var(k)+w[k];
  }
  map f(j)=Rr(j+i),p(1..n);
  ideal JJ=f(j)(K(j));
  if(dim(std(JJ))>i)
  {
   execute("ring A(j)=(complex,16,I),("+varstr(S)+"),dp;");
   list W(i)=imap(TJ(0),Ww(i));
   list w(j)=var(1);
  }
  else
  {
   if(i==0)
   {
    execute("ring RR(j)=(complex,16,I),("+varstr(S)+"),dp;");
    list W(i)=imap(TJ(0),Ww(i));
    list w=W(i)[j];
    ideal J=imap(TJ(0),N(0));
    def AA(j)=JuReZeroDim( J,w,d);
    setring AA(j);
    list W(i)=imap(TJ(0),Ww(i));
    if(t<1)
    {
     execute("ring A(j)=(complex,16,I),("+varstr(S)+"),dp;");
     list W(i)=imap(TJ(0),Ww(i));
     list w(j)=W(i)[j];
    }
    else
```

```
{
execute("ring RRR(j)=(complex,106,I),("+varstr(S)+"),dp;");
list W(i)=imap(TJ(0),Ww(i));
list w=W(i)[j];
ideal J=imap(TJ(0),N(0));
def AAA(j)=JuReTopDim( J,w, z(d),d);
setring AAA(j);
number ts=t;
list W(i)=imap(TJ(0),Ww(i));
if(ts<1)
{
 dt=d-1;
}
else
{
 dt=d;
}
if(dt>i)
{
 for(ii=i+1;ii<=dt;ii++)
 {
  execute("ring RRRR(ii+j)=(complex,106,I),("+varstr(S)+"),ds;");
  list Ww(ii)=imap(TJ(0),Ww(ii));
  if(size(Ww(ii)[1])>1)
  {
   execute("ring RRRRR(ii+j)=(complex,16,I),("+varstr(S)+"),dp;");
   list W(i)=imap(TJ(0),Ww(i));
   list w=W(i)[j];
   list Ww(ii)=imap(TJ(0),Ww(ii));
   ideal J=imap(TJ(0),N(0));
   ideal L=imap(TJ(0),LL);
   ideal L(ii);
   for(k=1;k<=ii;k++)
   {
   L(ii)[k]=L[k];
   }
   def AAA(ii+j)=ReJunkUseHomo(J,L(ii),Ww(ii),w);
   setring AAA(ii+j);
   number ts=t;
   list W(i)=imap(TJ(0),Ww(i));
   if(ts>0)
   {
   execute("ring A(j)=(complex,16,I),("+varstr(S)+"),dp;");
   list W(i)=imap(TJ(0),Ww(i));
   list w(j)=var(1);
   ii=d+1;
   }
   else
   {
```

```
      if(ii==dt)
      {
       execute("ring A(j)=(complex,16,I),("+varstr(S)+"),dp;");
       list W(i)=imap(TJ(0),Ww(i));
      }
      list w(j)=W(i)[j];
     }
    }
   }
  }
  else
  {
   execute("ring A(j)=(complex,16,I),("+varstr(S)+"),dp;");
   list W(i)=imap(TJ(0),Ww(i));
   list w(j)=W(i)[j];
  }
 }
}
else
{
 execute("ring RRRRRRR(j)=(complex,106,I),("+varstr(S)+"),dp;");
 list W(i)=imap(TJ(0),Ww(i));
 list w=W(i)[j];
 ideal J=imap(TJ(0),N(0));
 def Aaa(j)=JuReTopDim( J,w,z(d),d);
 setring Aaa(j);
 list W(i)=imap(TJ(0),Ww(i));
 number ts =t;
 if(ts<1)
 {
  dt=d-1;
 }
 else
 {
  dt=d;
 }
 if(dt>i)
 {
  for(ii=i+1;ii<=dt;ii++)
  {
   execute("ring RRRRRRRR(ii+j)=(complex,106,I),("+varstr(S)+"),
                                                   ds;");
   list Ww(ii)=imap(TJ(0),Ww(ii));
   if(size(Ww(ii)[1])>1)
   {
    execute("ring R1(ii+j)=(complex,16,I),("+varstr(S)+"),dp;");
    list W(i)=imap(TJ(0),Ww(i));
    list w=W(i)[j];
    list Ww(ii)=imap(TJ(0),Ww(ii));
```

```
      ideal J=imap(TJ(0),N(0));
      ideal L=imap(TJ(0),LL);
      ideal L(ii);
      for(k=1;k<=ii;k++)
      {
       L(ii)[k]=L[k];
      }
      def AA(ii+j)=ReJunkUseHomo(J,L(ii),Ww(ii),w);
      setring AA(ii+j);
      number ts=t;
      list W(i)=imap(TJ(0),Ww(i));
      if(ts>0)
      {
       execute("ring A(j)=(complex,16,I),("+varstr(S)+"),dp;");
       list W(i)=imap(TJ(0),Ww(i));
       list w(j)=var(1);
       ii=d+1;
      }
      else
      {
       if(ii==dt)
       {
       execute("ring A(j)=(complex,16,I),("+varstr(S)+"),dp;");
       list W(i)=imap(TJ(0),Ww(i));
       }
       list w(j)=W(i)[j];
      }
     }
    }
   }
   else
   {
    execute("ring A(j)=(complex,16,I),("+varstr(S)+"),dp;");
    list W(i)=imap(TJ(0),Ww(i));
    list w(j)=W(i)[j];
   }
  }
 }
if(j==z(i))
{
 execute("ring R(i)=(complex,16,I),("+varstr(S)+"),dp;");
 list W(i),w;
 int k(i)=0;
 for(k=1;k<=z(i);k++)
 {
  w=imap(A(k),w(k));
  if(size(w)>1)
  {
   k(i)=k(i)+1;
```

```
      W(i)[k(i)]=w;
     }
    }
    if(k(i)==0)
    {
     W(i)=var(1);
    }
   }
  }
 }
}
execute("ring T=(complex,16,I),("+varstr(S)+"),dp;");
int bt=0;
for(i=0;i<=d-1;i++)
{
 list Ww(i)=imap(TJ(0),W(i));
 if(size(Ww(i)[1])>1)
 {
  list W(i)=imap(R(i),W(i));
 }
 else
 {
  list W(i)=Ww(i);
 }
 export(W(i));
}
list W(d)=imap(TJ(0),W(d));
export(W(d));
ideal L=imap(TJ(0),LL);
export(L);
ideal N(0)=imap(TJ(0),N(0));
export(N(0));
setring S;
return(T);
 }
}
example
{ "EXAMPLE:";echo = 2;
   ring r=0,(x,y,z),dp;
   poly f1=(x3+z)*(x2-y);
   poly f2=(x3+y)*(x2-z);
   poly f3=(x3+y)*(x3+z)*(z2-y);
   ideal I=f1,f2,f3;
   def W=WitSet(I);
   setring W;
   W(1);
    // witness point  set of a pure 1-dimensional component of V(I)
   W(0);
   // witness point  set of a pure 0-dimensional component of V(I)
```

```
     L;
     // list of generic linear polynomials
}
//////////////////////////////////////////////////////////////////////////
static proc ZSR1(ideal I, ideal L, list W )
"USAGE:  ZSR1(ideal I, ideal L, list W );I ideal,
         L ideal defined by generic linear polynomials,
         W list of a point in the generic slicing of V(I) and V(L)
RETURN:  ts number;zero sum relation of W
EXAMPLE: example ZSR1;shows an example
"
{
 def S=basering;
 int n=nvars(basering);
 int c(1)=5*n;
 int c(2)=23*n;
 int iii=size(L);
 execute("ring R=(complex,16,I),("+varstr(S)+"),ds;");
 number c(0);
 ideal LL=imap(S,L);
 c(0)=leadcoef(LL[iii]);
 string sv=varstr(S);
 int j,ii,jj,k,a,b,te,zi,si;
 string sf,sff;
 list VV;
 list W=imap(S,W);
 VV[1]=W;
 def SB1=Singular2betini1(VV);
 execute("ring R=(complex,16,I),("+varstr(S)+",gamma,s),dp;");
 ideal I=imap(S,I);
 zi=size(I);
 ideal LL=imap(S,L);
 ideal H, ll;
 for(a=1;a<=zi;a++)
 {
  H[a]=s*gamma*I[a]+(1-s)*I[a];
 }
 if(iii>1)
 {
  for(k=1;k<=iii-1;k++)
  {
   ll[k]=LL[k];
   H[k+zi]=s*gamma*LL[k]+(1-s)*ll[k];
  }
  ll[iii]=LL[iii]+c(1)-c(0);
  H[iii+zi]=s*gamma*LL[iii]+(1-s)*ll[iii];
 }
 else
 {
```

```
 ll[iii]=LL[iii]+c(1)-c(0);
  H[iii+zi]=s*gamma*LL[iii]+(1-s)*ll[iii];
}
def Q(1)=UseBertini(H,sv);
string siaa=read("singular_solutions");
string saa=read("nonsingular_solutions");
def TT(1)=bertini2Singular("nonsingular_solutions",nvars(basering)-2);
setring TT(1);
list wr=re;
if(size(wr)==0)
{
 execute("ring TT(2)=(complex,16,I),("+varstr(S)+"),dp;");
 number tte, ts;
 tte=11;
 ts=0;
 export(ts);
 export(tte);
}
else
{
 execute("ring R1=(complex,16,I),("+varstr(S)+",gamma,s),dp;");
 ideal I=imap(S,I);
 si=size(I);
 ideal LL=imap(S,L);
 ideal H, ll;
 for(a=1;a<=si;a++)
 {
  H[a]=s*gamma*I[a]+(1-s)*I[a];
 }
 if(iii>1)
 {
  for(k=1;k<=iii-1;k++)
  {
   ll[k]=LL[k];
   H[k+si]=s*gamma*LL[k]+(1-s)*ll[k];
  }
  ll[iii]=LL[iii]+c(2)-c(0);
  H[iii+si]=s*gamma*LL[iii]+(1-s)*ll[iii];
 }
 else
 {
  ll[iii]=LL[iii]+c(2)-c(0);
  H[iii+si]=s*gamma*LL[iii]+(1-s)*ll[iii];
 }
 def Q(2)=UseBertini(H,sv);
 string saaa=read("nonsingular_solutions");
 string siaaa=read("singular_solutions");
 if(size(saaa)<52)
 {
```

```
  if(size(siaaa)<52)
  {
   "ERROR( Try again try);";
  }
 }
 if(size(saaa)>=52)
 {
  def TT(2)=bertini2Singular("nonsingular_solutions",nvars(basering)-2);
  setring TT(2);
  list wwr=re;
  list wr=imap(TT(1),wr);
  list W=imap(S,W);
  list w,ww,www;
  number s(0),s(1),s(2),ts;
  zi=size(W)/n;
  for(jj=1;jj<=zi;jj++)
  {
   s(0)=0;
   s(1)=0;
   s(2)=0;
   w=W;
   ww=wr[jj];
   www=wwr[jj];
   for(j=1;j<=n;j++)
   {
    s(0)=s(0)+j*w[j];
   }
   for(j=1;j<=n;j++)
   {
    s(1)=s(1)+j*ww[j];
   }
   for(j=1;j<=n;j++)
   {
    s(2)=s(2)+j*www[j];
   }
  }
  ts=s(0)*(c(1)-c(2))+s(1)*(c(2)-c(0))+s(2)*(c(0)-c(1));
 }
 else
 {
  def TT(2)=bertini2Singular("singular_solutions",nvars(basering)-2);
  setring TT(2);
  list wwr=re;
  list wr=imap(TT(1),wr);
  list W=imap(S,W);
  list w,ww,www;
  number s(0),s(1),s(2),ts;
  zi=size(W)/n;
  for(jj=1;jj<=zi;jj++)
```

```
  {
   s(0)=0;
   s(1)=0;
   s(2)=0;
   w=W;
   ww=wr[jj];
   www=wwr[jj];
   for(j=1;j<=n;j++)
   {
    s(0)=s(0)+j*w[j];
   }
   for(j=1;j<=n;j++)
   {
    s(1)=s(1)+j*ww[j];
   }
   for(j=1;j<=n;j++)
   {
    s(2)=s(2)+j*www[j];
   }
   }
  ts=s(0)*(c(1)-c(2))+s(1)*(c(2)-c(0))+s(2)*(c(0)-c(1));
 }
}
execute("ring e=(complex,16,I),("+varstr(S)+"),dp;");
number ts=imap(TT(2),ts);
export(ts);
number tte;
tte=11;
export(tte);
system("sh","rm start");
setring S;
return (e);
}
example
{ "EXAMPLE:";echo = 2;
    ring r=(complex,16,I),(x,y,z),dp;
    poly f1=(x2+y2+z2-6)*(x-y)*(x-1);
    poly f2=(x2+y2+z2-6)*(x-z)*(y-2);
    poly f3=(x2+y2+z2-6)*(x-y)*(x-z)*(z-3);
    ideal J=f1,f2,f3;
    ideal L=2*x+7*y+3*z+29;
    list W=2,1.999999999999999,-15.6666666666664;
    def D=ZSR1(J,L,W );
    setring D;
    ts;
}
//////////////////////////////////////////////////////////////////
static proc perSumZ(list A)
"USAGE:  perSumZ(list A);A list of different complex numbers
```

```
RETURN:   all subsets of A, whose sum of their elements is zero
EXAMPLE: example perSumZ;shows an example
"
{
 list B, C;
 int i, j;
 number t,tr;
 if(size(A)==0)
 {
  B[1]=A;
 }
 if(size(A)==1)
 {
  if(((repart(A[1]))^2+(impart(A[1]))^2)<=0.000000000000001)
  {
   B[1]=A;
  }
 }
 for(i=1;i<=size(A);i++)
 {
  t=t+A[i];
 }
 if(((repart(t))^2+(impart(t))^2)<=0.000000000000001)
 {
  B[1]=A;
 }
 for(i=1;i<=size(A);i++)
 {
  C=delete(A,i);
  C=perSumZ(C);
  for(j=1;j<=size(C);j++)
  {
   if(size(C[j])>0)
   {
    B[size(B)+1]=C[j];
   }
  }
 }
 return(B);
}
example
{ "EXAMPLE:";echo = 2;
   ring r=(complex,16,I),x,lp;
   list A=1,-1,2-I,I,-2;
   def D=perSumZ(A);
   D;
}
///////////////////////////////////////////////////////////////////
static proc  ZSROFWitSet(ideal I)
```

```
"USAGE:   ZSROFWitSet(ideal I);I ideal
RETURN:   ZSR(i) lists of the zero sum relation of witness point
           sets W(i) for i=1,...dim(V(I))
EXAMPLE: example ZSROFWitSet;shows an example
"
{
 def S=basering;
 int n=nvars(basering);
 def T(0)=WitSet(I);
 setring T(0);
 ideal LL=L;
 int dd=size(LL);
 int a=c(0);
 if(a==0)
 {
  return(T(0));
 }
 else
 {
  int i,j,ii,jj,k,sv(0..dd),j(0..dd),kk;
  string sv;
  for(i=1;i<=dd;i++)
  {
   jj=0;
   list V(i)=imap(T(0),W(i));
   if(size(V(i)[1])>1)
   {
    if(size(V(i))==1)
    {
     execute("ring L(i)(1)=(complex,16,I),("+varstr(S)+"),dp;");
     list W(i),ZSR(i);
     list V(i)=imap(T(0),W(i));
     W(i)=V(i);
     ZSR(i)[1]=0.000;
    }
    else
    {
     if(i>1)
     {
      execute("ring ee(i)=(complex,16,I),("+varstr(S)+"),dp;");
      list V(i)=imap(T(0),W(i));
     }
     sv(i)=size(V(i));
     for(j=1;j<=sv(i);j++)
     {
      ideal N=imap(T(0),N(0));
      ideal LLL=imap(T(0),LL);
      ideal L;
      for(kk=1;kk<=i;kk++)
```

```
      {
       L[kk]=LLL[kk];
       }
      def L(i)(j)=ZSR1(N,L,V(i)[j]);
      setring L(i)(j);
      if(j==1)
      {
       list W(i),ZSR(i);
       }
      else
      {
       list W(i)=imap(L(i)(j-1),W(i));
       list ZSR(i)=imap(L(i)(j-1),ZSR(i));
       export(ZSR(i));
       }
      list V(i)=imap(T(0),W(i));
      jj=jj+1;
      ZSR(i)[jj]=ts;
      W(i)[jj]=V(i)[j];
     }
    }
   }
  }
execute("ring Q=(complex,12,I),("+varstr(S)+"),dp;");
list W(0)=imap(T(0),W(0));
export(W(0));
for(jj=1;jj<=dd;jj++)
{
 number pt(jj);
 list V(jj)=imap(T(0),W(jj));
 if(size(V(jj)[1])>1)
 {
  sv(jj)=size(V(jj));
  if(jj>0)
  {
   list ZSR(jj)=imap(L(jj)(sv(jj)),ZSR(jj));
   export(ZSR(jj));
   list W(jj)=imap(L(jj)(sv(jj)),W(jj));
   export(W(jj));
  }
 }
 else
 {
  list ZSR(jj)=var(1);
  export(ZSR(jj));
  list W(jj)=var(1);
  export(W(jj));
 }
}
```

```
    ideal L=imap(T(0),LL);
    export(L);
    export(dd);
    system("sh","rm singular_solutions");
    system("sh","rm nonsingular_solutions");
    system("sh","rm real_solutions");
    system("sh","rm raw_solutions");
    system("sh","rm raw_data");
    system("sh","rm output");
    system("sh","rm midpath_data");
    system("sh","rm main_data");
    system("sh","rm input");
    system("sh","rm failed_paths");
    setring S;
    return(Q);
  }
}
example
{ "EXAMPLE:";echo = 2;
    ring  r = 0,(x,y,z),dp;
    poly f1=(x2+y2+z2-6)*(x-y)*(x-1);
    poly f2=(x2+y2+z2-6)*(x-z)*(y-2);
    poly f3=(x2+y2+z2-6)*(x-y)*(x-z)*(z-3);
    ideal J=f1,f2,f3;
    def D=ZSROFWitSet(J);
    setring D;
    ZSR(1);
    W(1);
    ZSR(2);
    W(2);
}
//////////////////////////////////////////////////////////////////////////
static proc ReWitZSR(list A, list W, int di)
"USAGE:  ReWitZSR(list A, list W, int di); A ideal of complex numbers,
          W list of points on di-dimensional component,
          di integer
RETURN:  tw(di) integer, list Z(size(Z));
         if tw(di)>0, else Z(0), list Z1(tw1(di))
EXAMPLE: example ReWitZSR;shows an example
"
{
 def S=basering;
 execute("ring e=(complex,16,I),("+varstr(S)+"),dp;");
 list A=imap(S,A);
 list W=imap(S,W);
 list  D, B(1..size(A)),C(1..size(A)),D(1..size(A)),Z1(1..size(A)),
       Z(1..size(A)),Z,Y,ZY,ZZ;
 int i,j,k,tw(di),tw1(di),tw2(di),tw3(di),tr,tc;
 list AA;
```

```
list WW;
for(i=1;i<=size(A);i++)
{
 if(((repart(A[i]))^2+(impart(A[i]))^2)<=0.0000000000000001)
 {
  tc=tc+1;
  tw1(di)=tw1(di)+1;
  ZY[i]=A[i];
  Z1(tw1(di))=W[i];
  export(Z1(tw1(di)));
 }
 else
 {
  tr=tr+1;
  tw2(di)=tw2(di)+1;
  AA[tr]=A[i];
  WW[tr]=W[i];
 }
}
A=AA;
W=WW;
if(size(A)>0)
{
 def B=perSumZ(A);
 for(i=1;i<=size(A);i++)
 {
  tc=0;
  B(i)=A[i];
  for(j=1;j<=size(B);j++)
  {
   tr=0;
   for(k=1;k<=size(B[j]);k++)
   {
    if(B(i)[1]==B[j][k])
    {
     tr=tr+1;
    }
   }
   if(tr>0)
   {
    tc=tc+1;
    C(i)[tc]=B[j];
   }
  }
  for(j=1;j<=size(C(i));j++)
  {
   D(i)=C(i)[j];
   for(k=1;k<=size(C(i));k++)
   {
```

```
      if(size(D(i))<size(C(i)[k]))
      {
       D(i)=D(i);
      }
      else
      {
       D(i)=C(i)[k];
      }
     }
    }
   }
   for(i=1;i<=size(A);i++)
   {
    Z[i]=D(i);
   }
   for(i=1;i<=size(Z);i++)
   {
    if(size(Z[i])>0)
    {
     D=Z[i];
     for(k=size(Z);k>0;k--)
     {
      if(size(Z[k])>0)
      {
       B=Z[k];
       if(i!=k)
       {
        if(D[1]==B[1])
        {
         Z=delete(Z,k);
        }
       }
      }
     }
    }
   }
   for(j=1;j<=size(Z);j++)
   {
      tr=0;
      D=Z[j];
      for(i=1;i<=size(A);i++)
      {
       for(k=1;k<=size(D);k++)
       {
        if(A[i]==D[k])
        {
         tr=tr+1;
         tw(di)=tw(di)+1;
         Z(j)[tr]=W[i];
```

```
      }
     }
    }
    export(Z(j));
  }
 export(Z);
 }
 if(tw1(di)==0)
 {
  list Z1(0);
  Z1(0)="Empty set";
  export(Z1(0));
 }
 if(tw(di)==0)
 {
  list Z(0);
  Z(0)="Empty set";
  export(Z(0));
 }
 export(tw1(di));
 export(tw(di));
  setring S;
  return (e);
}
example
{ "EXAMPLE:";echo = 2;
    ring r=(complex,16,I),(x,y,z),dp;
    list A= 3.7794571034732007+I*21.1724850800421247,
          -3.7794571034752664-I*21.1724850800419908;
    list W=list(-2.0738016397747976,1.29520655909919,
                                  -0.1476032795495952),
          list(-1.354769788796631,-1.5809208448134761,
                                  1.2904604224067381);
    int di=1;
    def D=ReWitZSR(A,W,di);
    setring D;
    tw(di);
    Z(size(Z));// if tw(di)>0, else Z(0);
    Z1(tw1(di));
}
///////////////////////////////////////////////////////////////////////
proc NumIrrDecom(ideal I) Numerical Irreducible Decomposition
"USAGE:  NumIrrDecom(ideal I);I ideal
RETURN:  w(1),..., w(t) lists of irreducible witness point sets of
         irreducible components of V(J)
EXAMPLE: example NumIrrDecom;shows an example
"
{
 def S=basering;
```

```
int i,ii;
def WW=ZSROFWitSet(I);
setring WW;
if(c(0)==0)
{
 setring S;
 return(WW);
}
else
{
 int d=size(L);
 for(i=0;i<=d;i++)
 {
  int co(i)=0;
  if(i==0)
  {
   execute("ring q(i)=(complex,16,I),("+varstr(S)+"),dp;");
   list V(i)=imap(WW,W(i));
   list W(0..size(V(i)));
   if(size(V(i)[1])>1)
   {
    co(i)=size(V(i));
    for(ii=1;ii<=size(V(i));ii++)
    {
     list w(ii)=V(i)[ii];
     export(w(ii));
    }
   }
   else
   {
    W(1)[1]="Empty Set";
   }
  }
  else
  {
   list WW(i);
   list V(i)=imap(WW,W(i));
   list a(i)=imap(WW,ZSR(i));
   if(size(V(i)[1])>1)
   {
    def q(i)=ReWitZSR(a(i),V(i),i);
    setring q(i);
    if(tw1(i)>0)
    {
     for(ii=1;ii<=tw1(i);ii++)
     {
      WW(i)[ii]=Z1(ii);
     }
     co(i)=tw1(i);
```

```
    }
    if(tw(i)>0)
    {
     for(ii=1;ii<=size(Z);ii++)
     {
      if(size(Z[ii])>1)
      {
       co(i)=co(i)+1;
       WW(i)[ii+tw1(i)]=Z(ii);
      }
     }
    }
    for(ii=1;ii<=size(WW(i));ii++)
    {
     list w(ii);
     w(ii)=WW(i)[ii];
    }
   }
   else
   {
    execute("ring q(i)=(complex,16,I),("+varstr(S)+"),dp;");
    WW(i)[1]="Empty Set";
   }
  }
 }
 for(i=0;i<=d;i++)
 {
  execute("ring qq(i)=(complex,16,I),("+varstr(S)+"),dp;");
  for(ii=1;ii<=co(i);ii++)
  {
   list w(ii)=imap(q(i),w(ii));
   export w(ii);
  }
  "=========================================";
  "=========================================";
  "Dimension";
   i;
  "Number of Components";
   co(i);
  number cco(i)=co(i)/1;
  export(cco(i));
 }
 ideal L=imap(WW,L);
 export(L);
 "The generic Linear Space L";
 L;
 return(qq(0..d));
 }
}
```

```
example
{ "EXAMPLE:";echo = 2;
    ring r=0,(x,y,z),dp;
    poly f1=(x2+y2+z2-6)*(x-y)*(x-1);
    poly f2=(x2+y2+z2-6)*(x-z)*(y-2);
    poly f3=(x2+y2+z2-6)*(x-y)*(x-z)*(z-3);
    ideal I=f1,f2,f3;
    list W=NumIrrDecom(I);
      ==>
         Dimension
         0
         Number of Components
         1
         Dimension
         1
         Number of Components
         3
         Dimension
         2
         Number of Components
         1
    def A(0)=W[1];
      // corresponding 0-dimensional components
    setring A(0);
    w(1);
        // corresponding 0-dimensional irreducible component
        ==> 0-Witness point set (one point)
    def A(1)=W[2];
            // corresponding 1-dimensional components
    setring A(1);
    w(1);
        // corresponding 1-dimensional irreducible component
        ==> 1-Witness point set (one point)
    w(2);
      // corresponding 1-dimensional irreducible component
        ==> 1-Witness point set (one point)
    w(3);
      // corresponding 1-dimensional irreducible component
        ==> 1-Witness point set (one point)
    def A(2)=W[3];
    // corresponding 2-dimensional components
    setring A(2);
    w(1);
      // corresponding 2-dimensional irreducible component
        ==> 1-Witness point set (two points)
}
//////////////////////////////////////////////////////////////////
proc defl(ideal I, int d)
"USAGE:   defl(ideal I, int d);  I ideal, int d order of the deflation
```

```
RETURN:    deflation ideal DI of I
EXAMPLE:   example defl; shows an example
"
{
 def S=basering;
 int n=nvars(basering);
 int i,j;
 for(i=1;i<=d;i++)
 {
  def R(i)=symmetricBasis(n,i,"x");
  setring R(i);
  ideal J(i)=symBasis;
  export(J(i));
 }
 execute("ring RR=0,(x(1..n),"+varstr(S)+"),dp;");
 for(i=1;i<=d;i++)
 {
  ideal J(i)=imap(R(i),J(i));
  for(j=1;j<=n;j++)
  {
   J(i)=subst(J(i),x(j),var(n+j));
  }
 }
 execute("ring R=0,("+varstr(S)+"),dp;");
 ideal I=imap(S,I);
 if(d>1)
 {
  for(i=1;i<=d-1;i++)
  {
   ideal J(i)=imap(RR,J(i));
   for(j=1;j<=size(I);j++)
   {
    ideal I(j);
    for(k=1;k<=size(J(i));k++)
    {
     I(j)[k]=J(i)[k]*I[j];
    }
     export(I(j));
   }
  }
  ideal J(d)=imap(RR,J(d));
  ideal D(d)=J(1..d);
  ideal II(d)=I,I(1..size(I));
  matrix T(d)=diff(D(d),II(d));
  matrix TT(d)=transpose(T(d));
  export(TT(d));
 }
 else
 {
```

```
 ideal J(d)=imap(RR,J(d));
 ideal D(d)=J(d);
 ideal II(d)=I;
 matrix T(d)=diff(D(d),II(d));
 matrix TT(d)=transpose(T(d));
 export(TT(d));
}
int zc=size(D(d));
export(zc);
execute("ring DR=0,("+varstr(S)+",x(1..zc)),dp;");
matrix TT(d)=imap(R,TT(d));
ideal I=imap(S,I);
vector v=[x(1..zc)];
ideal DI=I,TT(d)*v;
export(DI);
export(I);
setring S;
return(DR);
}
example
{ "EXAMPLE:"; echo = 2;
  ring r=0,(x,y,z),dp;
  poly f1=z^2;
  poly f2=z*(x^2+y);
  ideal I=f1,f2;
  def D=defl(I,1);
  setring D;
  DI;
}
//////////////////////////////////////////////////////////////////////
static proc    NIDofDI(ideal I)
"USAGE:  NIDofDI(ideal I);  I ideal
RETURN:  numerical irreducible decomposition of I
EXAMPLE: NIDofDI; shows an example
"
{
 def S=basering;
 int i,ii;
 def WW=ZSROFWitSet(I);
 setring WW;
 if(c(0)==0)
 {
  setring S;
  return(WW);
 }
 else
 {
  int d=size(L);
  for(i=0;i<=d;i++)
```

```
{
 int co(i)=0;
 if(i==0)
 {
  execute("ring q(i)=(complex,16,I),("+varstr(S)+"),dp;");
  list V(i)=imap(WW,W(i));
  list W(0..size(V(i)));
  if(size(V(i)[1])>1)
  {
   co(i)=size(V(i));
   for(ii=1;ii<=size(V(i));ii++)
   {
    list w(ii)=V(i)[ii];
    export(w(ii));
   }
  }
  else
  {
   W(1)[1]="Empty Set";
  }
 }
 else
 {
  list WW(i);
  list V(i)=imap(WW,W(i));
  list a(i)=imap(WW,ZSR(i));
  if(size(V(i)[1])>1)
  {
   def q(i)=ReWitZSR(a(i),V(i),i);
   setring q(i);
   if(tw1(i)>0)
   {
    for(ii=1;ii<=tw1(i);ii++)
    {
     WW(i)[ii]=Z1(ii);
    }
    co(i)=tw1(i);
   }
   if(tw(i)>0)
   {
    for(ii=1;ii<=size(Z);ii++)
    {
     if(size(Z[ii])>1)
     {
      co(i)=co(i)+1;
      WW(i)[ii+tw1(i)]=Z(ii);
     }
    }
   }
```

```
    for(ii=1;ii<=size(WW(i));ii++)
    {
     list w(ii);
     w(ii)=WW(i)[ii];
     }
    }
    else
    {
     execute("ring q(i)=(complex,16,I),("+varstr(S)+"),dp;");
     WW(i)[1]="Empty Set";
    }
   }
  }
  for(i=0;i<=d;i++)
  {
   execute("ring qq(i)=(complex,16,I),("+varstr(S)+"),dp;");
   list ww(i);
   if(co(i)>0)
   {
    for(ii=1;ii<=co(i);ii++)
    {
    list v(ii)=imap(q(i),w(ii));
    ww(i)=v(ii)[1];
    if(size(ww(i))==1)
    {
     list w(ii);
     w(ii)[1]=v(ii);
    }
    else
    {
     list w(ii)=v(ii);
    }
    export(w(ii));
    }
   }
   else
   {
    list w(1);
    w(1)[1]=var(1);
    export(w(1));
   }
   number cco(i)=co(i)/1;
   export(cco(i));
  }
  ideal L=imap(WW,L);
  export(L);
  return(qq(0..d));
 }
}
```

```
example
{ "EXAMPLE:"; echo = 2;
   ring r=0,(x,y,z),dp;
   poly f1=z^2;
   poly f2=z*(x^2+y);
   ideal I=f1,f2;
   list DD=NIDofDI(I);
   def D(0)=DD[1];
   setring D(0);
   w(1);           // w(1)= x, i.e. no components
   def D(1)=DD[2];
   setring D(1);
   w(1);
   def D(2)=DD[3];
   setring D(2);
   w(1);
}
//////////////////////////////////////////////////////////////////////////
proc NumPrimDecom(ideal I,int d)
"USAGE:    NumPrimDecom(ideal I,int d); I ideal, d order of the deflation
RETURN:    lists of the numerical primary decomposition
EXAMPLE:   example NumPrimDecom; shows an example
"
{
 def S=basering;
 int n=nvars(basering);
 int i,Dd,j,k,jj;
 def D=defl(I,d);
 setring D;
 ideal J=DI;
 Dd=dim(std(DI));
 list W=NIDofDI(J);
 for(i=0;i<=Dd;i++)
 {
  def A(i+1)=W[i+1];
  setring A(i+1);
  if(cco(i)>0)
  {
   for(j=1;j<=cco(i);j++)
   {
    list W(j)=w(j);
    for(k=1;k<=size(w(j));k++)
    {
     for(jj=size(W(j)[k]);jj>=n+1;jj--)
     {
      W(j)[k]=delete(W(j)[k],jj);
     }
     W(j)[k]=W(j)[k];
    }
```

```
   }
  }
  else
  {
   list W(1)=var(1);
  }
 }
 execute("ring R=(complex,16,I),("+varstr(S)+"),dp;");
 jj=0;
 for(i=0;i<=Dd;i++)
 {
  number cco(i)=imap(A(i+1),cco(i));
  if(cco(i)>0)
  {
   for(j=1;j<=cco(i);j++)
   {
    jj=jj+1;
    list w(jj)=imap(A(i+1),W(j));
    export(w(jj));
 "=========================================";
 "=========================================";
    "Numerical Primary Component";
    w(jj);
   }
  }
 }
 return(R);
}
example
{ "EXAMPLE:"; echo = 2;
  ring r=0,(x,y),dp;
  poly f1=yx;
  poly f2=x2;
  ideal I=f1,f2;
  def W=NumPrimDecom(I,1);
  setring W;
  w(1);
     ==> 1-Witness point set (one point)
  w(2);
     ==> 1-Witness point set (one point)
}

//////////////////////////////////////////////////////////////////
//////////////////////////////////////////////////////////////////
```

5.2 Library "NumerAlg.lib"

```
LIBRARY:  NumerAlg.lib    Numerical Algebraic Algorithm
OVERVIEW:
    The library contains procedures to test the inclusion and the equality
    of two ideals defined by polynomial systems, compute the degree of a
    pure i-dimensional component of an algebraic variety defined by a pol-
    ynomial system, compute the local dimension of an algebraic variety
    defined by a polynomial system at a point computed as an approximate
    value. The use of the library requires to install Bertini
                    (www.nd.edu/~sommese/bertini/thedownload)
PROCEDURES:

 Incl(ideal I, ideal J);   test if I containes J

 Equal(ideal I, ideal J);  test if I equals to J

 Degree(ideal I, int i);   computes the degree of a pure i-dimensional

 NumLocalDim(ideal I, p);  computes local dimension numerically at a point computed as
                           an approximate value

LIB "NumerDecom.lib";

//////////////////////////////////////////////////////////////////////////
//////////////////////////////////////////////////////////////////////////
proc Degree(ideal I,int i)
"USAGE:  Degree(ideal I,int i); I ideal,  i positive integer
RETURN:  the degree of the pure i-dimensional component of the algebraic
         variety defined by I
NOTE:    if the degree is -1 , i.e, no components of dimension i
EXAMPLE: example Degree; shows an example
"
{
 def S=basering;
 def W=WitSet(I);
 setring W;
 int j;
 if(size(W(i)[1])>1)
 {
  j=size(W(i));
 }
 else
 {
  j=-1;
 }
 "The Degree of Component";
 j;
 setring S;
```

```
  return (W);
}
example
{ "EXAMPLE:"; echo = 2;
   ring r=0,(x,y,z),dp;
   poly f1=(x2+y2+z2-6)*(x-y)*(x-1);
   poly f2=(x2+y2+z2-6)*(x-z)*(y-2);
   poly f3=(x2+y2+z2-6)*(x-y)*(x-z)*(z-3);
   ideal I=f1,f2,f3;
   def W=Degree(I,1);
       ==>
          The Degree of Component
          3
   def W=Degree(I,2);
       ==>
          The Degree of Component
          2
}
/////////////////////////////////////////////////////////////////////////
proc Incl(ideal I, ideal J)
"USAGE:  Incl(ideal I, ideal J); I, J ideals
RETURN:  t=1 if the algebraic variety defined by I contains the algebraic
          variety defined by J, otherwise t=0
EXAMPLE: example Incl; shows an example
"
{
 def S=basering;
 int n=nvars(basering);
 int i,j,ii,k,z,zi,dd;
 if(dim(std(I))==0)
  {
   def W=solve(I,"nodisplay");
   setring W;
   ideal J=imap(S,J);
   ideal I=imap(S,I);
   list w;
   poly tj;
   number al,ar,ai,ri,jj;
   zi=size(SOL);
   for(j=1;j<=zi;j++)
    {
     w=SOL[j];
     for(k=1;k<=size(J);k++)
      {
       tj=J[k];
       for(ii=1;ii<=n;ii++)
        {
         tj=subst(tj,var(ii),w[ii]);
        }
```

```
   al=leadcoef(tj);
   ar=repart(al);
   ai=impart(al);
   ri=ar^2+ai^2;
   if(ri>0.000000000000001)
   {
     jj=0;
     k=size(I)+1;
     j=zi+1;
   }
   else
   {
     jj=1;
     ri=0;
   }
  }
 }
}
else
{
 def W=WitSupSet(I);
 setring W;
 ideal J=imap(S,J);
 ideal I=imap(S,I);
 list w;
 number al,ar,ai,ri,jj;
 poly tj;
 dd=size(L);
 for(i=0;i<=dd;i++)
 {
  z=size(W(i)[1]);
  zi=size(W(i));
  if(z>1)
  {
   for(j=1;j<=zi;j++)
   {
    w=W(i)[j];
    for(k=1;k<=size(J);k++)
    {
     tj=J[k];
     for(ii=1;ii<=n;ii++)
     {
      tj=subst(tj,var(ii),w[ii]);
     }
     al=leadcoef(tj);
     ar=repart(al);
     ai=impart(al);
     ri=ar^2+ai^2;
     if(ri>0.000000000000001)
```

```
       {
        jj=-1;
        k=size(J)+1;
        j=zi+1;
        z=0;
        i=dd+1;
        }
        else
        {
        jj=1;
        ri=0;
        }
       }
      }
     }
    }
   }
 }
 if(ri>0.000000000000001)
 {
  jj=0;
 }
 else
 {
  jj=1;
 }
"===============================================";
 "Inclusion:";
 jj;
"===============================================";
 export(jj);
 export(J);
 export(I);
   system("sh","rm singular_solutions");
   system("sh","rm nonsingular_solutions");
   system("sh","rm real_solutions");
   system("sh","rm raw_solutions");
   system("sh","rm raw_data");
   system("sh","rm output");
   system("sh","rm midpath_data");
   system("sh","rm main_data");
   system("sh","rm input");
   system("sh","rm failed_paths");
 setring S;
 return (W);
}
example
{ "EXAMPLE:"; echo = 2;
   ring r=0,(x,y,z),dp;
   poly f1=(x2+y2+z2-6)*(x-y)*(x-1);
```

```
    poly f2=(x2+y2+z2-6)*(x-z)*(y-2);
    poly f3=(x2+y2+z2-6)*(x-y)*(x-z)*(z-3);
    ideal I=f1,f2,f3;
    poly g1=(x2+y2+z2-6)*(x-1);
    poly g2=(x2+y2+z2-6)*(y-2);
    poly g3=(x2+y2+z2-6)*(z-3);
    ideal J=g1,g2,g3;
    def W=Incl(I,J);
       ==>
          Inclusion:
          0
    def W=Incl(J,I);
       ==>
          Inclusion:
          1
}
/////////////////////////////////////////////////////////////////////////
proc Equal(ideal I, ideal J)
"USAGE: Equal(ideal I, ideal J); I, J ideals
RETURN: t=1 if the algebraic variety defined by I equals to the algebraic
           variety defined by J, otherwise t=0
EXAMPLE: example Equal; shows an example
"
{
 def S=basering;
 int n=nvars(basering);
 def W1=Incl(J,I);
 setring W1;
 number j1=jj;
 execute("ring q=(real,0),("+varstr(S)+"),dp;");
 ideal I=imap(W1,I);
 ideal J=imap(W1,J);
 execute("ring qq=0,("+varstr(S)+"),dp;");
 ideal I=imap(S,I);
 ideal J=imap(S,J);
 def W2=Incl(I,J);
 setring W2;
 number j2=jj;
 number j;
 number j1=imap(W1,j1);
 if(j2==1)
 {
  if(j1==1)
  {
   j=1/1;
  }
  else
  {
   j=0/1;
```

```
 }
}
else
{
 j=0/1;
}
"===============================================";
"Equality:";
j;
"===============================================";
setring S;
return (W2);
}
example
{ "EXAMPLE:"; echo = 2;
  ring r=0,(x,y,z),dp;
  poly f1=(x2+y2+z2-6)*(x-y)*(x-1);
  poly f2=(x2+y2+z2-6)*(x-z)*(y-2);
  poly f3=(x2+y2+z2-6)*(x-y)*(x-z)*(z-3);
  ideal I=f1,f2,f3;
  poly g1=(x2+y2+z2-6)*(x-1);
  poly g2=(x2+y2+z2-6)*(y-2);
  poly g3=(x2+y2+z2-6)*(z-3);
  ideal J=g1,g2,g3;
  def W=Equal(I,J);
      ==>
          Equality:
          0
  def W=Equal(J,J);
      ==>
          Equality:
          1
}
////////////////////////////////////////////////////////////////////////
proc NumLocalDim(ideal J, list w, int e)
"USAGE:  NumLocalDim(ideal J, list w, int e); J ideal,
          w list of an approximate value of a point v in the algebraic
          variety defined by J, e integer
RETURN: the local dimension of the algebraic variety defined by J at v
EXAMPLE: example NumLocalDim; shows an example
"
{
 def S=basering;
 int n=nvars(basering);
 int sI=size(J);
 int i,j,jj,t,tt,sz1,sz2,ii,ph,ci,k;
 poly p,pp;
 list rw,iw;
 for(i=1;i<=sI;i++)
```

```
{
 p=J[i];
 for(j=1;j<=n;j++)
 {
  w[j]=w[j]+I*0;
  rw[j]=repart(w[j]);
  iw[j]=impart(w[j]);
  p=subst(p,var(j),w[j]);
 }
 pp=pp+p;
}
number u=leadcoef(pp);
if((u^2)==0)
{
 execute("ring A=(real,e-1),("+varstr(S)+",I),ds;");
 ideal II=imap(S,J);
 list rw=imap(S,rw);
 list iw=imap(S,iw);
 poly p(1..n);
 for(j=1;j<=n;j++)
 {
  p(j)=var(j)+rw[j]+I*iw[j];
 }
 map f=A,p(1..n);
 ideal T=f(II);
 tt=dim(std(T));
 t=tt-1;
}
else
{
 int d=dim(std(J));
 execute("ring R=(complex,e-1,I),("+varstr(S)+"),ds;");
 list w=imap(S,w);
 ideal II=imap(S,J);
 ideal JJ;
 poly p, p(1..n);
 for(i=1;i<=sI;i++)
 {
  p=II[i];
  for(j=1;j<=n;j++)
  {
   p=subst(p,var(j),w[j]);
  }
  JJ[i]=II[i]-p;
 }
 for(j=1;j<=n;j++)
 {
  p(j)=var(j)+w[j];
 }
```

```
map f=R,p(1..n);
ideal T=f(JJ);
tt=dim(std(T));
if(tt==d)
{
 execute("ring A=(complex,e,I),("+varstr(S)+"),dp;");
 t=tt;
}
else
{
 execute("ring RR=(real,e-2),("+varstr(S)+",I),dp;");
 ideal II=imap(S,J);
 list rw=imap(S,rw);
 list iw=imap(S,iw);
 ideal L,LL,H,HH;
 poly l(1..d),ll(1..d);
 int c;
 for(i=1;i<=d;i++)
 {
  for(j=1;j<=n;j++)
  {
   c=random(1,100);
   l(i)=l(i)+c*(var(j));
   ll(i)=ll(i)+c*(var(j)-rw[j]-I*iw[j]);
  }
  l(i)=l(i)+random(101,200);
  L[i]=l(i);
  LL[i]=ll(i);
 }
 poly pi=I^2+1;
 H=L,II,pi;
 ideal JJ;
 poly p, p(1..n);
 for(i=1;i<=sI;i++)
 {
  p=II[i];
  for(j=1;j<=n;j++)
  {
   p=subst(p,var(j),rw[j]+I*iw[j]);
  }
  JJ[i]=II[i]-p;
 }
 HH=LL,JJ,pi;
 if(dim(std(H))==0)
 {
  def M=solve(H,100,"nodisplay");
  setring M;
  sz1=size(SOL);
  execute("ring RRRQ=(real,e-1),("+varstr(S)+",I),dp;");
```

```
ideal HH=imap(RR,HH);
if(dim(std(HH))==0)
{
 def MM=solve(HH,100,"nodisplay");
 setring MM;
 sz2=size(SOL);
}
}
else
{
 sz1=1;
}
if(sz1==sz2)
{
 execute("ring A=(complex,e,I),("+varstr(S)+"),dp;");
 t=d;
}
else
{
 execute("ring RQ=(real,e-1),("+varstr(S)+"),dp;");
 ideal II=imap(S,J);
 def RW=WitSet(II);
 setring RW;
 list v;
 list w=imap(S,w);
 number nr,ni;
 if(tt<0)
 {
  tt=0;
 }
 for(ii=tt;ii<=d;ii++)
 {
  list W(ii)=imap(RW,W(ii));
  if(size(W(ii)[1])>1)
  {
   if(ii==0)
   {
    for(i=1;i<=size(W(0));i++)
    {
     v=W(ii)[i];
     nr=0;
     ni=0;
     for(j=1;j<=n;j++)
     {
      nr=nr+(repart(v[j])-repart(w[j]))^2;
      ni=ni+(impart(v[j])-impart(w[j]))^2;
     }
     if((ni+nr)<1/10^(2*e-3))
     {
```

```
    execute("ring A=(complex,e,I),("+varstr(S)+"),dp;");
    list W(ii)=imap(RW,W(ii));
    t=0;
    i=size(W(ii))+1;
    ii=d+1;
   }
  }
}
else
{
 def SS=Singular2betini1(W(ii));
 execute("ring D=(complex,e,I),("+varstr(S)+",s,gamma),dp;");
 string nonsin;
 ideal H,L;
 ideal J=imap(RW,N(0));
 ideal LL=imap(RW,L);
 list w=imap(S,w);
 poly p;
 for(j=1;j<=ii;j++)
 {
  p=0;
  for(jj=1;jj<=n;jj++)
  {
   p=p+random(1,100)*(var(jj)-w[jj]);
  }
  L[j]=p;
 }
 for(jj=1;jj<=size(J);jj++)
 {
  H[jj]=s*gamma*J[jj]+(1-s)*J[jj];
 }
 for(jj=1;jj<=ii;jj++)
 {
  H[size(J)+jj]=s*gamma*LL[jj]+(1-s)*L[jj];
 }
 string sv=varstr(S);
 def Q(ii)=UseBertini(H,sv);
 system("sh","rm start");
 nonsin=read("nonsingular_solutions");
 if(size(nonsin)>=52)
 {
  def T(ii)=bertini2Singular("nonsingular_solutions",
                                        nvars(basering)-2);
  setring T(ii);
  list C=re;
  ci=size(C);
  number tr;
  list w=imap(S,w);
  for(jj=1;jj<=ci;jj++)
```

```
            {
            tr=0;
            for(k=1;k<=n;k++)
            {
              tr=tr+(repart(w[k])-repart(C[jj][k]))^2+(impart(w[k])-
                                                impart(C[jj][k]))^2;
            }
            if(tr<=1/10^(2*e-3))
            {
              execute("ring A=(complex,e,I),("+varstr(S)+"),dp;");
              t=ii;
              ii=d+1;
              jj=ci+1;
            }
          }
        }
      }
    }
  }
  system("sh","rm singular_solutions");
  system("sh","rm nonsingular_solutions");
  system("sh","rm real_solutions");
  system("sh","rm raw_solutions");
  system("sh","rm raw_data");
  system("sh","rm output");
  system("sh","rm midpath_data");
  system("sh","rm main_data");
  system("sh","rm input");
  system("sh","rm failed_paths");
  }
 }
}
"=========================================";
"The Local Dimension:";
t;
setring S;
return(A);
}
example
{ "EXAMPLE:"; echo = 2;
    int e=14;
    ring r=(complex,e,I),(x,y,z),dp;
    poly f1=(x2+y2+z2-6)*(x-y)*(x-1);
    poly f2=(x2+y2+z2-6)*(x-z)*(y-2);
    poly f3=(x2+y2+z2-6)*(x-y)*(x-z)*(z-3);
    ideal J=f1,f2,f3;
    list p0=0.99999999999999+I*0.00000000000001,2,3+I*0.00000000000001;
    list p2=1,0.99999999999998,2;
    list p1=5+I,4.999999999999998+I,5+I;
```

```
    def D=NumLocalDim(J,p0,e);
            ==>
                The Local Dimension:
                0
    def D=NumLocalDim(J,p1,e);
            ==>
                The Local Dimension:
                1
    def D=NumLocalDim(J,p2,e);
            ==>
                The Local Dimension:
                2
}
```

///
///

Bibliography

[1] E.L. Allgower and K. Georg. *Numerical Continuation Methods, an Introduction*, Springer Series in Comput. Math, Vol. 13, Springer- Veralg, Berlin, Heidelberg, New Yourk, 1990.

[2] S. Al Rashed and G. Pfister. *Numerical Decomposition of Affine Algebraic Varieties*. Preprint, 2011.

[3] E. Arbarello, M. Coranalba, P.A. Griffths and J. Harris. *Geometry of Algebraic Curves*, volume I. Volume 267 of Grundlehren Math. Wiss., Springer Verlag, New York, 1985.

[4] D.J. Bates, J.D. Hauenstein, A.J. Sommese and C.W. Wampler . BERTINI : *Software for Numerical Algebraic Geometry*. *http://www.nd.edu/~sommese/bertini/*.

[5] D.J. Bates, J.D. Hauenstein, C. Peterson and A.J. Sommese . *Numerical Local Dimension Test for Points on the Solution Set of a System of Polynomial Equations*. SIAM J. Number. Anal. Vol. 47, No. 5, pp. 3608 - 3623. Novemebr 13, 2009.

[6] S.C. Billups, A.P. Morgan and L.T. Watson . Algorithm 652. Hompack: *A Suite codes for Globally Convergent Homotopy Algorithms*. ACM Transactions on Mathematical Software, Vol. 13. No. 3, Septemper 1987, Pages 281 - 310.

[7] G. Björck and R. Fröberg. *A faster way to count the solutions of inhomogeneous systems of algebraic equations, with applications to cyclic n-roots*. J. Symbolic Computation, 12: 329 - 336, 1991.

[8] G. Björck. *Functions of modulus one on Z_n whose Fourier transforms have constant modulus, and cyclic n-roots*. In: J.S. Byrnes and J.F. Byrnes, Editors, Recent Advances in Fourier Analysis and its Applications 315, NATO Adv. Sci. Inst. Ser. C. Math. Phys. Sci., Kluwer (1989), pp. 131 - 140.

[9] Chow, Mallet-Paret, and Yorke: *Total degree homotopy*, 1978.

[10] R. M. Corless, A. Galligo, I. S. Kotsireas and S. M. Watt. *A Geometric-Numeric Algorithm for Absolute Factorization of Multivariate Polynomials*. ISSAC 2002, July 7 - 10, 2002, Lille, France.

[11] B. Dayton and Z. Zeng . *Computing the Multiplicity Structure in Solving Polynomial System*, in Proceedings of ISSAC 2005, ACM, New Yourk, 2005, pp. 116 - 123.

[12] J. Dennis, J. Robert, B. Schnabel: *Numerical Methods for Unconstrained Optimization and Nonlinear Equations*. 1996 Society for Industrial and Applied Mathematics.

[13] W. Decker, G.-M Greuel, G. Pfister and H. Schönemann. Singular 3-1-1 — A computer algebra system for polynomial computations. http://www.singular.uni-kl.de (2010).

[14] W. Decker, G.-M. Greuel, G. Pfister: *"Primary decomposition: algorithms and comparisons."* In: G.-M. Greuel, B.H. Matzat, G. Hiss: Algorithmic Algebra and Number Theory. Springer Verlag, Heidelberg (1998), 187 - 220.

[15] G. Fischer. *Complex Analytic Geometry*, Lecture Notes in Mathematics 538 (1976).

[16] W. Fulton. *Intersection Theory*, volume (3) 2 of *Ergeb. Math. Grenzgeb.* Springer Verlag, Berlin, 1984.

[17] G.-M. Greuel and G. Pfister. A Singular *Introduction to Commutative Algebra*. Second edition, Springer (2007).

[18] A. Leykin: *Numerical Primary Decomposition*. ISSAC'08, July 20-23, 2008, Hagenberg, Austria. Copyright 2008 ACM 978 - 1 - 59593 - 904 - 3/08/07.

[19] A. Leykin, J. Verschelde and A. Zhao. *Higher-order deflation for polynomial systems with isolated singular solutions*. In A. Dickenstein, F.-O. Schreyer and A.J. Sommese, editors, Algorithms in Algebraic Geometry, volume 146 of The IMA Volumes in Mathematics and its Applications. Springer, 2008.

[20] A. Leykin, J. Verschelde and A. Zhao. *Newtons method with deflation forisolated singularities of polynomial systems*. Theoretical Computer Science, $359(1-3): 111-122$, 2006.

[21] T.Y. Li. *Numerical solutions of multivariate polynomial systems by homotopy continuation methods.* Acta Number. 6 (1997), 399 - 436.

[22] A.P. Morgan and L.T. Watson. *A globally convergent parallel algorithm for zeros of polynomial systems.* Nonlinear Anal. (1989), 13(11), 1339 - 1350.

[23] A.P. Morgan and A.J. Sommese. *Computing all solutions to polynomial systems using homotopy continuation.* Appl. Math. Comput., 115 - 138. Errata: Appl. Math. Comput. 51 (1992), p. 209. Nonlinear Anal.

[24] Morgan's Book on polynomial continuation 1987.

[25] A.J. Sommese, J. Verschelde and C.W. Wampler. *Symmetric functions applied to decomposing solution sets of polynomial systems.* Vol. 40, No. 6, pp. 2026 - 2046. 2002 Society for Industrial and Applied Mathematics.

[26] A.J. Sommese and J. Verschelde. *Numerical homotopies to compute generic points on positive dimensional algebraic sets. J. of Complexity 16(3):572 - 602, (2000).*

[27] A.J. Sommese, J. Verschelde and C.W. Wampler. (2002a). *A method for tracking singular paths with application to the numerical decomposition* . In algebraic geometry (pp. 329 - 345). Berlin: de Gruyter.

[28] A.J. Sommese, J. Verschelde and C.W. Wampler. (2001a). *Numerical decomposition of solution sets of polynomial systems into irreducible components.* SIAM J. Number. Anal., 38(6), 2022 - 2046.

[29] A.J. Sommese, J. Verschelde and C.W. Wampler. (2003). *Numerical irreducible decomposition using PHCpack.* In algebra, geometry, and software systems (pp. 109- 129). Berlin: Springer.

[30] A.J. Sommese, J. Verschelde and C.W. Wampler. *Numerical decomposition of the solution sets of polynomial systems into irreducible components.* SIAM J. Numer. Anal. 38(6):2022 - 2046, 2001.

[31] A.J. Sommese, J. Verschelde and C.W. Wampler. *Numerical irreducible decomposition using projections from points on components.* In Symbolic Computation: Solving Equations in Algebra, Geometry, and Engineering, volume 286 of contemporary Mathematics, edited by E.L. Green, S. Hosten, R,C. Laubenbacher, and V. Powers, pages 27 - 51. AMS 2001.

[32] A.J. Sommese, J. Verschelde and C.W. Wampler. *Using monodromy to decompose solution sets of polynomial systems into irreducible components*. In Application of Algebraic Geometry to Coding Theory, Physics and Computation, edited by C. Ciliberto , F. Hirzebruch, R. Miranda, and M. Teicher. Proceedings of a NATO Conference, February 25-March 1, 2001, Eilat, Israel. Pages 297 - 315, Kluwer Academic Publishers.

[33] A.J. Sommese and C.W. Wampler. *The Numerical Solution of Systems of Polynomials Arising in Engineering and Science*. ISBN 981 - 256 - 184 - 6. Word Scientific Publishing Co. Plte. Ltd. 2005.

[34] A.J. Sommese and C.W. Wampler. *Numerical algebraic geometry. In the mathematics of numerical analysis*. (Park City, UT, 1995), Vol. 32 of lectures in Appl. Math. (pp. 749 - 763). Providence, RI: Amer. Math. Soc.

[35] Hai-Jun Su, J. Micheal Mc Carthy, and Layne. T. Watson: *Algorithm 857: PolySYS-GLP-A Parallel General Linear Product Homotopy Code for Solving Polynomial Systems of Equations*. ACM Transaction on Mathematical Software, Vol. 32, No. 4, December 2006, Pages 561 - 579.

[36] J. Verschelde(1996). *Homotopy continuation methods for solving polynomial systems*. PhD thesis, Katholihe Universiteit Leuven.

[37] J. Verschelde(1999). Algorithm 795: PHCpack:*Ageneral-purpose solver for polynomial systems by homotopy continuation*. ACM Trans. on Math. Software, 25(2), 251 - 276.

[38] Layne.T. Watson: *Globally Convergent Homotopy Algorithms for Nonlinear Systems of Equation*. Nonlinear Dynamics 1:143 - 191. 1990 Kluwer Academic Publishers.

Die VDM Verlagsdienstleistungsgesellschaft sucht für unsere
Redaktion engagierte Mitarbeiter...nnen und herausragende

Dissertationen, Habilitationen,
Diplomarbeiten, Master-Thesen,
Magisterarbeiten usw.

Ihr eigenes Buch - kostenlose Publikation als Fachbuch

Wir legen Wert auf einen reibungslosen Ablauf der inhaltlichen und professionellen Betreuung der Autoren. Interessierte an einer kostenlosen Buchpublikation

Bringen Sie Ihre akademische Abschlussarbeit zur bekannten Geltung!

Sie erhalten kostenlos ein Exemplar Ihres Buches!

VDM Verlagsdienstleistungsgesellschaft mbH
Dudweiler Landstr. 99   Telefon: +49 681 3720 174
D-66123 Saarbrücken   +49 681 3720 109

www.vdm-vsg.de